Stars Over Hawai'i

by
E.H. Bryan, Jr.

2002 and 2006 Additions by
Richard A. Crowe, Ph.D.
2015 Revisions by
Timothy F. Slater, Ph.D.

PETROGLYPH
PRESS

Royalties on the sale of *Stars Over Hawai'i* benefit The Bernice Pauahi Bishop Museum and the Dr. Richard Crowe Memorial Scholarship at the University of Hawaii at Hilo. Donations are accepted through the University of Hawaii Foundation, 200 West Kawili St., Hilo, Hawaii 96720 or online at:

https://giving.uhfoundation.org/#funds/20623903
or http://jasminecrowemusic.com/scholarship.

Star Compasses on pages 63-66 have been included with permission from Polynesian Voyaging Society. The Polynesian Voyaging Society is a not-for-profit organization. Contributions towards their efforts to teach the traditional art of Polynesian voyaging and their Malama Honua Worldwide Voyage may be sent to Ala Moana Blvd., Pier 7, Honolulu, Hawai'i 96813.

NASA Space Photos
Front Cover: NASA/ESA - Sagittarius dwarf irregular galaxy
Back Cover: NASA/Thomas M. Brown/Randy A. Kimble/Allen V. - Ultraviolet light source in an old galaxy

Constellation figures drawn by E.M. Brownlee.
Front and back cover illustrations and spot illustrations throughout book are by Dietrich Varez.
All other drawings by E.H. Bryan, Jr.

Māui the Fisherman & Hina Releasing the Moon and Stars *by Dietrich Varez*

The constellation Kamakaunuiomāui (Māui's Fishhook, also known as Scorpius) appears prominently in the skies over Hawai'i. According to legend, Māui used his fishhook to pull the Hawaiian Islands up from the sea floor. As the islands arose from the sea, Māui's fishhook flew up to the sky and became the constellation we see today. Both the legend and the constellation are depicted on the front cover. The artwork on the back cover represents the legend of the goddess Hina releasing the moon and stars from her calabash as they flew up to take their place in the skies. Other block illustrations appear throughout the book. Dietrich Varez prints are available from The Volcano Art Center, The Honolulu Museum of Art, Kokee Museum, Bishop Museum Gift Shop and DietrichVarez.com.

Published by Petroglyph Press, Ltd.
1672 Kamehameha Avenue • Hilo, Hawai'i 96720 • USA
Voice 808-935-6006 • Fax 808-935-1553
PetroglyphPress@hawaiiantel.net
www.PetroglyphPress.com • Facebook.com/PetroglyphHilo

Contents

Publishers Note

Originally compiled by the late Edwin Bryan, Jr. in 1955, *Stars Over Hawai'i* celebrates 60 consecutive years in print with the 2015 release of this new edition. It is geared to presenting the night sky as seen over Hawai'i to further generations of backyard observers in an accessible format. In 2002 the late Dr. Richard Crowe, Professor of Astronomy and Physics at the University of Hawaii at Hilo, embarked on an update that contributed a tremendous amount of new information covering the previous 25 years. Current theories about the formation of our solar system, new knowledge gained through space exploration and powerful modern telescopes, and a genuine and heartfelt respect for Hawaiian culture and history fueled his additions and greatly improved the content. He contributed the chapter on Polynesian Voyaging and Wayfinding and facilitated the addition of the Star Compasses developed by the Polynesian Voyaging Society and Nainoa Thompson. Dr. Crowe's colleague the late Dr. Walter Steiger, Professor Emeritus of Astronomy and Physics at UH Manoa, was a friend and contemporary of Ed Bryan while at Bishop Museum. He assisted with the update and wrote Bryan's biography. This latest edition further updates current theories and astronomical knowledge courtesy of Dr. Timothy Slater, Astronomer at the University of Wyoming in Laramie and a senior scientist at the CAPER Center for Astronomy & Physics Education Research. *Stars Over Hawai'i* is largely historical and, although much effort has been made to update scientific theories and cultural information, some of the cultural references contained within reflect the period in which it was originally written.

Publishers Acknowledgements

The publishers gratefully acknowledge the many hours of research and writing that the late Dr. Richard Crowe invested in making the long overdue additions and revisions to *Stars Over Hawai'i* for the 2002 edition, the first major update in 25 years. The results far exceeded what we imagined when we first approached him with the project. Richard was not one to do things half way and his loss is deeply felt. We extend our sincerest appreciation to the Polynesian Voyaging Society for permission to reprint the Star Compasses. Mahalo to Dr. Tim Slater who, at a chance meeting at HawaiiCon, offered his time and effort to bring this new edition up to date. Stacey Reed contributed sharp proofreading, thorough attention to small details, and vital suggestions.

Mahalo to Dietrich Varez for kindly providing us with the illustrations shown throughout the book, and for the use of his color illustrations on the front and back covers.

With the collaboration of these people and others *Stars Over Hawai'i* has become a work of which we are truly proud.

David and Christine Reed, July, 2015

A Note from Dr. Richard Crowe – 2002

The completion of this book would not have been possible without the contributions of several key individuals. A big mahalo to Master Navigator Chad Kalepa Baybayan for his review of Chapters 2, 8, and 9, for providing valuable input about the star compasses as well as the Hawaiian gourd compass, and for contributing Ka Heihei o na Keiki. Mahalo also to Dennis Kawaharada for providing high-quality reproductions of the star compasses. We would like to acknowledge the Polynesian Voyaging Society for permission to use the star compasses and accompanying descriptions. We are deeply grateful to Walt Steiger for his assistance with rewriting Chapter 6, and for providing an updated version of the Lahaina Noon statistics. Thanks also to Michael West of Maria Mitchell Association formerly of UH Hilo and Gemini Observatory for his critical review of Chapters 4 and 5. Some material in Chapters 4 and 5 was referenced from *Understanding the Cosmos*, by Richard A. Crowe (1994, Whittier, New York, NY); some material in Chapters 2, 8, and 9 was referenced from *Na Inoa Hoku*, by Rubellite K. Johnson and John K. Mahelona (1975, Topgallant Publishing, Honolulu, HI).

Chapter 1
Stars Over Hawai'i

Hawai'i is a grand place from which to view the stars. Nights are mild, skies are generally clear; best of all, from Hawai'i you can see all of the bright stars and most of the important constellations in the sky at some time during the year.

Hawai'i is located at about 20 degrees north latitude near the northern edge of the tropics. The Pole Star Polaris, which marks the northern celestial pole of the sky, stays at a convenient angle of about 20 degrees above the northern horizon. The angle of the northern pole of the sky above the northern horizon is always exactly equal to the north latitude of the observer. Throughout the year Polaris stands still while other stars appear to circle around it. All of the important star groups in the southern sky rise above the southern horizon at some time of year. Most of the 88 constellations can be seen clearly from Hawai'i. Only three are not seen at all. There are no bright stars located less than 20 degrees from the southern pole of the sky.

The chief object of this book is to help you to become acquainted with the stars and constellations as seen from Hawai'i. To do so we present twelve charts showing the stars as seen from the latitude of Hawai'i each month of the year. These charts are drawn and arranged here to show the heavens as seen in the early evening from January to December, but they can be made to show the position of stars at any hour of any night. The Star Chart Finder on page 18 shows which charts represent the heavens from sunset to sunrise throughout the year.

Other places around the world which have about the same north latitude as Hawai'i include: Wake Island, Saipan, Luzon, southern Formosa, Hainan Island, South China, Hanoi, Burma, Central India, Mecca (Saudi Arabia), the Upper Nile, the Sahara Desert, Puerto Rico, Haiti, Cuba, and Mexico City. These charts would be equally useful at all these places, for the stars would be visible in the same parts of the sky at about the same local time.

A second object of this book is to give a very simple introduction to the broad facts of astronomy. It does not pretend to be a textbook on the subject. There are numerous excellent textbooks on astronomy, but some people hesitate to read them because they are afraid that the subject will be too technical. Astronomy can be a very technical subject, but the main facts need not be too difficult to understand. Chapters 3, 4 and 5 try to outline the subject and explain some main facts about the Solar System, the stars, and galaxies. We hope that these brief statements will whet your appetite to want to learn more about astronomy.

Astronomy has many modern uses, such as, for example, the determination of position on the Earth, time and the calendar. Navigators of ships and airplanes use the stars and planets and the sun to tell where they are, just as the Polynesian navigators did and still do, although modern celestial navigation is augmented today by satellite signals. The explorer and the mapmaker know how they can pinpoint positions on the surface of the Earth by observing the heavenly bodies. Chapter 6, titled "When the Sun Casts No Shadow - Lahaina Noon," tells of events occurring each year in Hawai'i that cannot happen in the mainland United States, and illustrates the practical application of a knowledge of astronomy.

The Hawaiians of long ago made use of the apparent movement of the heavenly bodies across the sky as their clock and calendar, as well as to steer their canoes across the ocean. The arrival of familiar groups of stars at a certain time of night marked the season of the year for harvesting crops, going on trips, or holding special ceremonies. One glance at the sky and the expert fisherman knew that it was time to start for the reef or to push off in his canoe. Something will be said of these subjects in the last chapters.

People all over the world have found the objects in the sky useful and friendly, as should we, so let's get acquainted with the stars over Hawai'i.

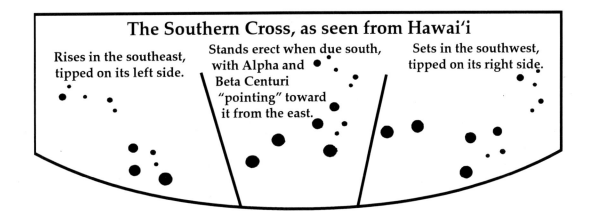

The Southern Cross, as seen from Hawai'i

Rises in the southeast, tipped on its left side.

Stands erect when due south, with Alpha and Beta Centuri "pointing" toward it from the east.

Sets in the southwest, tipped on its right side.

Chapter 2
The Constellations

Many centuries ago people looked up at the sky at night and thought that groups of stars formed figures. To these figures they gave names to honor characters or animals in their mythology. At first people in each region gave names of their own to the groups of stars. As years went by some of the names became well known in many lands, and were used by astronomers and astrologers to indicate exactly which stars they meant. For example, a person might say, "Aldebaran is the red star which marks the eye of the bull," or "Antares is the red star in the heart of the scorpion." The bull came to be known by its Latin name, Taurus; and the scorpion as Scorpio or Scorpius.

For many generations this information was handed down by word of mouth, just as the Hawaiians told their stories and passed along their knowledge. Then, about 130 C.E. (Common Era or A.D.) in Egypt, a writer of scientific information named Ptolemy collected the stories and made a catalog of star groups. Of the constellations that he listed, 48 are still recognized by modern astronomers and called by the same names. In more modern times, 40 other constellations have been added. Some of these are just "space fillers" between more prominent groups. Others are groups of stars in the Southern sky, too far south to be seen to advantage by these ancient astronomers, most of whom lived in Greece, Egypt, Persia, India and China. Some of the new constellations are named for modern mechanical objects and instruments,

such as an air pump, sculptor's chisel, pair of compasses, furnace, clock, microscope and sextant. This was because the European explorers who first charted the Southern hemisphere were seafarers. Their names stand out in contrast to the names of the older, more classical constellations, that reflect a hunter-gatherer society.

Of the 88 constellations now recognized, 30 are north of the zodiac, 12 make up the zodiac (a band of constellations through which the sun and the planets appear to move across the sky), and 46 are south of the zodiac. The names of the constellations are written in Latin, long the language of science. Many of the stars have individual names, some of them given by the Arabs. Other stars are called by Greek letters and are followed by the name of the constellations. When the stars were given these letter names, what seemed to be the brightest star in the constellation was called Alpha, the next brightest Beta, and so on. If we were to say "Alpha of Taurus" in Latin we would say "Alpha Tauri," using the genitive form of the constellation name. The genitive of some names is rather different from the nominative (regular) form of the name. For example, that of Crux (the southern cross) is Crucis. In the tabulation of constellations, that follows on pages 12 and 13, the genitive form and the meaning have been added. Also given is the name of the person (since Ptolemy) who named the constellation, the number of the star chart on which it is best shown, and whether it is in the zodiac or north or south of it.

Canis Major, Lepus, Orion, and Taurus

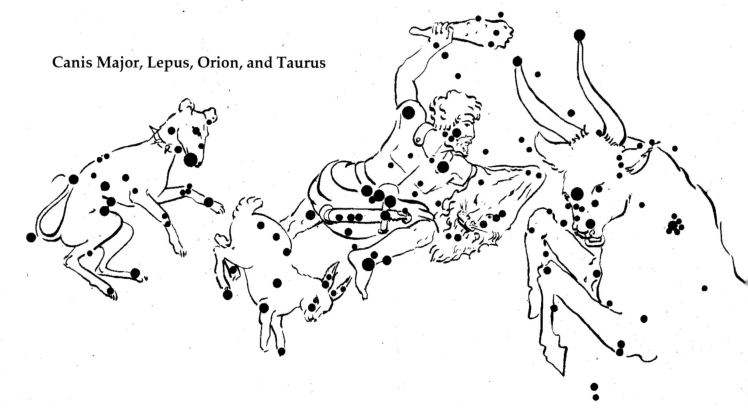

Some Constellation Stories

Most of the older constellations are supposed to represent, or at least are named after, characters or objects made famous by folk tales of long ago. Five constellations, most of them in the northern sky, are characters in a well-known classic myth. In the Greek account, Andromeda was the daughter of Cepheus, king of Ethiopia. His wife, Cassiopeia, a vain queen, boasted that she was more beautiful than the sea nymphs. This angered

Andromeda

Poseidon, god of the sea and father of the nymphs. He sent a sea monster, Cetus, to ravage Ethiopia. The Oracle of Amon told King Cepheus that, if he would save his country, he must sacrifice his beautiful daughter, so Andromeda was chained to a rock near the sea. Just as Cetus was about to devour Andromeda, along came Perseus, a warrior of the gods. He was returning from an adventure on which he had slain one of the wicked Gorgons, Medusa with the snakey locks. The head of Medusa, which Perseus had in his bag, had the power of turning into stone anything that looked at it. Perseus pulled out the head, turned Cetus to stone, rescued Andromeda, and after other adventures, he married her. All of the main characters in the story are present in the

sky. Some are quite lifelike, shown in such detail that even the bag containing the head of Medusa can be seen hanging from the belt of Perseus, Medusa's winking eye being represented by the variable star Algol (actually an eclipsing binary), which is first bright, then faint.

Another famous pair of constellations in the northern sky are Ursa Major, the greater bear, and Ursa Minor, the lesser bear. They are being driven 'round and

Perseus

'round the northern pole of the sky by Bootes, the herdsman or bear driver. The "Big Dipper" represents the hindquarters and tail of the greater bear. The "Little Dipper" is the hindquarters and tail of the lesser bear with Polaris, the pole star, at the end of its tail. Draco, the dragon, twists its long body between the two bears. (See chart N, which shows the stars near the northern pole of the sky, on page 31).

Stories about other northern groups, such as Hercules, the famous giant; Auriga, the chariot driver; Lyra, the flying harp; Ophiuchus, a man holding a long snake in his hands; Pegasus, the winged horse; Aquila, the eagle; and Cygnus, the swan, the long neck of which forms the upright of the "northern cross," can be found

Cassiopeia and Cepheus

Cetus

Auriga

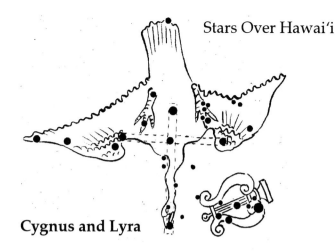

Cygnus and Lyra

in books of classic myths and other folklore, as well as in some astronomy books.

All the zodiac constellations date back to antiquity. Pisces represents the fishes of Venus and Cupid, which escaped a great typhoon by taking refuge in the river Euphrates. Aries is a ram, either the one into which Zeus (Jupiter) changed himself to escape some giants, or the

mous twins, Castor and Pollux, which are the names of its two brightest stars. Cancer, the crab, contains a noted cluster of stars, called Praesepe, the beehive. Leo is a lion, its head formed by a sickle-shaped curve of stars, with Regulus at the end of the handle of the sickle. Virgo is the virgin daughter of Jupiter; she is the goddess of justice. She holds Libra, the scales of justice. The stars in

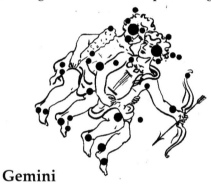

Gemini

Aquila

one with the Golden Fleece sought by Jason and the Argonauts. Taurus was a famous bull. Bulls were very highly regarded by the Chaldeans and other peoples of Asia. Its V-shaped face, below two long horns, is called the Hyades, with a bright red eye, Aldebaran. The little dipper-shaped Pleiades, or seven sisters, daughters of Atlas, are on the bull's back. Gemini represents two fa-

Libra formerly represented the claws of the scorpion, variously called Scorpio or Scorpius, which is the next zodiac group, with Antares marking its heart. Sagittarius is an archer, shooting an arrow into the scorpion. It also resembles a teapot with steam coming out of it (the Milky Way). Capricornus is a sea goat, with the head of a goat and the tail of a fish. Aquarius is a man pouring water from a jug.

Ophiuchus and Serpens

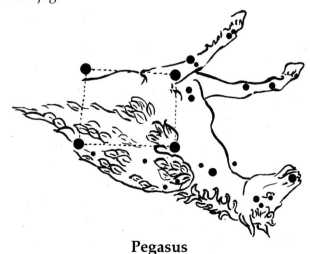

Pegasus

Aries

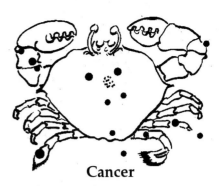

Cancer

The following are the most famous constellations south of the zodiac: Orion, a giant hunter, with belt and sword (outlined by the Great Nebula), has a club upraised in his right hand (the red supergiant Betelegeuse, marks his right shoulder) and a shield of lion skin outstretched in his left, ready to meet the charging bull. Orion is followed by his two faithful dogs - Canis Major, the great dog, and Canis Minor, the lesser dog, marked by the bright star Procyon. The brightest star in the sky, Sirius, is in the head or collar of the great dog, which is sniffing at Lepus, the realistic rabbit or hare. Argo, near the southern horizon in spring, as seen from Hawai'i, is the ship in which Jason and his companions went in search of the Golden Fleece, in a famous Greek story. This group is so large that a few centuries ago it was divided into four parts: Carina, the keel; Malus, the mast; Vela, the sails; and Puppis, the high stern. The rudder, beneath this last, is marked by Canopus, the second brightest star in the sky. Columba, the dove, is called Noah's dove by some Christian peoples. The great river Eridanus zigzags down the sky from near Rigel, in the left knee of Orion, to Achernar, a first magnitude star near Hawai'i's southern horizon in winter. Hydra, the snake, stretches a third of the way around the sky, with Sextans, Crater (the cup), and Corvus, the crow, astride its back. Centaurus, a creature that is half man and half horse, prances in the southern part of the Milky Way. Two of its bright stars, Alpha Centauri and Beta Centauri, seem to point from the east toward the famous Southern Cross (Crux), which lies between its legs. Most of the other southern constellations have modern names.

How to use the star charts

It is not hard to learn to identify the bright stars and constellation groups by using the star charts (pages 19 to 34). They are drawn to represent exactly the way the heavens appear from 20 degrees north latitude. Twelve of the charts show the positions of the stars at intervals

Aquarius and Piscis Australis

of a month at the same time of night. Since the stars appear to rise two hours earlier each month, as the Earth races along its orbit around the sun, two consecutive charts also represent a time difference of two hours for any particular night. Although the table of dates and hours in the lower left hand corner of each chart is for the early evening, the charts can be adapted to any time of night. The Star Chart Finder on Page 18 makes it easy to pick the chart that will represent the heavens at any time of night on any date.

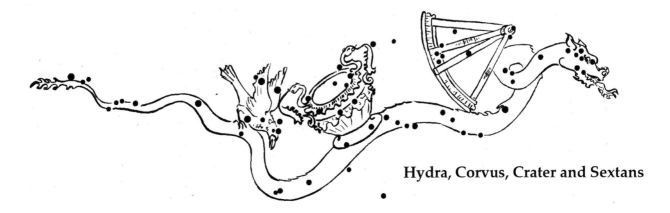

Hydra, Corvus, Crater and Sextans

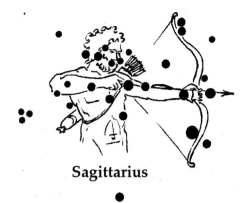

Sagittarius

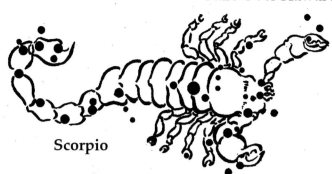

Scorpio

If you want to know when you can see a particular constellation to the best advantage, find what chart number follows the name of the constellation in the alphabetical list. On that chart it should be near the meridian or north-south line through the zenith or point above your head. A few constellations around the southern pole of the sky are not visible to advantage from Hawai'i and are not marked on the monthly charts. They are shown on an extra chart of the southern sky, Star Chart X: Southern Sky - South Polar Stars.

The "Mercator" star chart (pages 32 and 33) shows all of the constellations and brighter stars around the whole sky, but not as they would look at any particular time from Hawai'i. This chart illustrates the use of right ascension and declination to locate stars in the sky.

Orient yourself and your star chart

The charts are drawn to represent the heavens as you look up at them. You must hold the chart so that you look up at it also. When you face South (Hema), hold the chart upright in front of you, with its lower edge (marked "Southern Horizon") down, toward the south. If you face North ('Akau), tip the chart upside down and hold its upper ("Northern") edge toward the northern horizon. If you face West (Komohana), hold the right hand side of the chart (marked "Western Horizon") downward. If you face East (Hikina), hold the left hand side of the chart toward the eastern horizon.

Knowing the points of the compass will help you to identify the stars. Knowing the stars you can tell the points of the compass. East is the direction from which the stars, which later will be overhead, appear to rise. West is the direction toward which they will appear to set. From places north of the Equator such as Hawai'i, the North Star Polaris (Hokupa'a) is always visible in the northern sky. The north pole of the sky is at the same angle above the northern horizon as the north latitude of the observer. The pole star is only about one degree from this spot. You can locate the pole star by reference to the "pointers" of the Big Dipper, the open angles of the "W" of Cassiopeia, and the east and west sides of the Great Square of Pegasus, all of which indicate its direction. There is a special map on page 31 of the stars that seem to circle around the northern pole of the sky; these are called circumpolar stars.

Although the stars around the southern pole of the sky are not very bright, and are not visible from Hawai'i, there are various constellations and bright stars near our southern horizon that help to indicate the direction of south. These parade across the southern sky from east to west. Included among these are: the Southern Cross (Crux), Alpha and Beta Centauri, Canopus (in Argo), Achernar (at the end of the river Eridanus), and Fomalhaut (pronounced fo-mal-oh) in the southern fish.

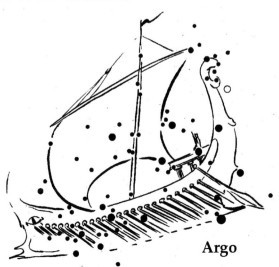

Argo

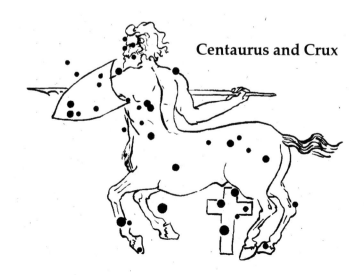

Centaurus and Crux

Leo

After you have oriented yourself, and are holding the correct chart in the proper way, you are ready to identify the stars. Start by matching up the bright stars and distinctive groups in the sky with the dots that represent them on the chart. This may not be easy at first, particularly if parts of the sky are obscured by clouds. But very soon you will be able to recognize even individual stars by their brightness and color, or by combinations of stars nearby, as easily as you can tell whole constellations.

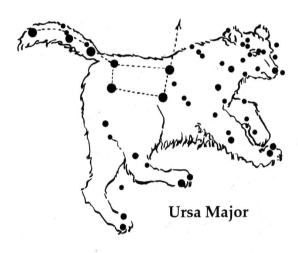

Ursa Major

In the summer you might begin with the Big Dipper, in the northern sky. It is really part of the greater bear (Ursa Major). The "pointers" at the western edge of the bowl of the dipper indicate the direction of Polaris, the North Star. The "handle" of the dipper or tail of the bear curves around in the direction of the bright star Arcturus (Hawai'i's Zenith Star), in Bootes, the herdsman or bear driver. From this star you can locate Regulus, to the west, at the base of the handle of the "sickle" of Leo, the lion. Spica, in Virgo, the virgin, lies to the southwest of Arcturus, with the trapezium shape of Corvus, the crow, beyond. If you look directly toward the southern horizon from Corvus (which resembles an upside-down milk pail) you will find the Southern Cross. The red supergiant Antares, the heart of the scorpion, is conspicuous in the southern sky. Vega, in Lyra, the flying harp, is in the northeast. Vega may be recognized by two slightly fainter stars that form with it an equilateral triangle. East

of Vega is Cygnus, the swan, with Deneb at the top of the group known as the Northern Cross. Southeast of Vega is Aquila, the eagle, containing the bright star Altair with a fainter star in line on each side. This trio of bright stars (Vega, Deneb, Altair) is also called the Summer, or Navigator's Triangle.

In winter the Pole Star is still in the same place in the northern sky, but all the other stars have shifted their position. The Big Dipper now is below the northern horizon, but the angles of "W"-shaped Cassiopeia open in the direction of Polaris. Toward the southwest is the "Great Square" of Pegasus, the eastern and western sides of which point almost due north and south. Altair, in Aquila, the eagle, is setting in the west. Orion, the mighty hunter, one of the most easily recognized constellations, is in the east, followed by Sirius, the "dog star" in Canis Major, the brightest star in the sky. Canopus, in the stern of the ship Argo, just rising far to the southeast, is the second brightest star. Taurus, the bull, is to the northwest of Orion, and Gemini, the twins, to its northeast. To the north of these is Auriga, the charioteer, containing the bright star Capella. Using these as a framework, it is easy to fill in and become acquainted with other stars and constellations.

Virgo

Table of Constellations

Name	Hawaiian Name	Genitive Form	Greek Meaning	Named By	Chart No.	Declination from Zodiac
Andromeda		Andromedae	[classic myth*]	P.	12	N
Antlia		Antliae	Air pump	Lac.	4	S
Apus		Apodis	Bird of Paradise	Bayer	X	S
Aquarius		Aquarii	Water Carrier	P.	11	Z
Aquila	Humuma	Aquilae	Eagle	P.	9	N
Ara		Arae	Altar	P.	8	S
Argo (Navis)		Argus	Argo (Ship)	P.	4	S
Aries	Hapakane	Arietis	Ram	P.	12	Z
Auriga	Hokolei	Aurigae	Charioteer	P.	2	N
Bootes	Hokuiwa	Bootis	Bear Keeper	P.	6	N
Caelum		Caeli	Sculptor's chisel	Lac.	X	S
Camelopardalis		Camelopardalis	Giraffe	Hev.	1	N
Cancer		Cancri	Crab	P.	3	Z
Canes Venatici		Canum Venaticorum	Hunting dogs	Hev.	6	N
Canis Major		Canis Majoris	Greater dog	P.	3	S
Canis Minor		Canis Minoris	Lesser dog	P.	3	S
Capricornus		Capricorni	Seagoat	P.	10	Z
Carina		Carinae	Keel (of ship Argo)	-	4	S
Cassiopeia	'Iwakeli'i	Cassiopeiae	[classic myth*]	P.	12	N
Centaurus		Centauri	Centaur	P.	6	S
Cepheus	Kamo'i	Cephei	[classic myth*]	P.	9	N
Cetus		Ceti	Sea monster*	P.	1	S
Chamaeleon		Chamaeleontis	Chameleon	Bayer	X	S
Circinus		Circini	Pair of compasses	Lac.	7	S
Columba		Columbae	Dove (Noah's dove)	–	2	S
Coma Berenices		Comae Berenices	Bernice's hair	T.B.	6	N
Corona Australis		Coronae Australis	Southern crown	P.	9	S
Corona Borealis	Kauamea	Coronae Borealis	Northern crown	P.	7	N
Corvus	Me'e	Corvi	Crow	P.	6	S
Crater		Crateris	Cup	P.	5	S
Crux	Hanaiakamalama	Crucis	Southern Cross	-	6	S
Cygnus		Cygni	Swan	P.	9	N
Delphinus	Nai'a	Delphini	Dolphin	P.	10	N
Dorado		Doradus	Swordfish	Bayer	1	S
Draco		Draconis	Dragon	P.	7	N
Equuleus		Equulei	Little horse	P.	10	N
Eridanus	Nu'uanu	Eridani	River Po (Italy)	P.	1	S
Fornax (Chemica)		Fornacis	(Chemical) Furnace	Lac.	1	S
Gemini	Namahoe	Geminorum	Twins	P.	3	Z
Grus		Grucis	Crane	Bayer	10	S
Hercules		Herculis	[classic myth*]	P.	8	N
Horologium		Horologii	Clock	Lac.	12	S
Hydra		Hydrae	Snake	P.	3-6	S
Hydrus		Hydri	Water-snake	Bayer	X	S
Indus		Indi	Indian	Bayer	10	S
Lacerta		Lacertae	Lizard	Hev.	11	N
Leo		Leonis	Lion	P.	5	Z
Leo Minor		Leonis Minoris	Lesser Lion	Hev.	5	N
Lepus		Leporis	Hare	P.	2	S
Libra		Librae	Balance, Scales	P.	7	Z

Name	Hawaiian Name	Genitive Form	Greek Meaning	Named By	Chart No.	Declination from Zodiac
Lupus		Lupi	Wolf	P.	7	S
Lynx		Lyncis	Lynx	Hev.	3	N
Lyra	Keho'oea	Lyrae	Harp	P.	9	N
Mensa (Mons Mensa)		Mensae	Table (Table Mountain)	Lac	X	S
Microscopium		Microscopii	Microscope	Lac.	10	S
Monoceros		Monocerotis	Unicorn	Hev.	3	N
Musca		Muscae	Fly	Bayer	X	S
Norma		Normae	Rule	Lac.	8	S
Octans		Octantis	Octant	Lac.	X	S
Ophiuchus		Ophiuchi	Serpent carrier	P.	8	N
Orion	Kaheiheionakeiki	Orionis	[classic myth*]	P.	2	S
Pavo		Pavonis	Peacock	Bayer	X	S
Pegasus	Kalupeokawelo	Pegasi	Winged horse	P.	11	N
Perseus		Persei	[classic myth*]	P.	1	N
Phoenix		Phoenicis	Phoenix	Bayer	12	S
Pictor		Pictoris	Painter's easel	Lac.	1	S
Pisces		Piscium	Fishes	P.	12	Z
Piscis Australis		Piscis Australis	Southern fish	P.	11	S
Puppis		Puppis	Stern (of ship Argo)	-	3	S
Reticulum		Reticuli	Net	Lac.	X	S
Sagitta		Sagittae	Arrow	P.	9	N
Sagittarius	Pimoe	Sagittarii	Archer	P.	9	Z
Scorpio	Kamakaunuiomaui	Scorpii	Scorpion	P.	8	Z
Sculptor		Sculptoris	Sculptor's apparatus	Lac.	12	S
Scutum (Sobeski)		Scuti	Shield (of Sobieski)	Hev.	9	N
Serpens		Serpentis	Serpent (snake)	P.	8	N
Sextans		Sextantis	Sextant	Hev.	5	S
Taurus		Tauri	Bull	P.	1	Z
Telescopium		Telescopii	Telescope	Lac.	9	S
Triangulum		Trianguli	Triangle	P.	1	N
Triangulum Australi		Trianguli Australis	Southern triangle	Bayer	8	S
Tucana		Tucanae	Toucan	Bayer	X	S
Ursa Major	Nahiku	Ursae Majoris	Greater Bear	P.	5	N
Ursa Minor		Ursae Minoris	Lesser Bear	P.	7	N
Vela		Velorum	Sails (of ship Argo)	~	4	S
Virgo		Virginis	Virgin	P.	6	Z
Volans		Volantis	Flying fish	Bayer	X	S
Vulpecula		Vulpeculae	Fox (and goose)	Hev.	9	N

P.   in Ptolemy's catalog
T.B.   Tycho Brahe (1546-1601)
Bayer   Johann Bayer (1572-1660)
Hev.   Johann Hevelius (1611-1687)
Lac.   Nicolas Louis de Lacaille (1713-1762)
Parts of names in parenthesis are usually omitted.

* See Constellation Stories for these classic myths.
N   North of Zodiac
Z   Zodiac constellation
S   South of Zodiac
X   Star Chart X: Southern Sky - South Polar Stars. Page 34. These constellations are not visible to advantage from Hawai'i.
Note: For Hawaiian Constellations and Star Names as they appear in the sky: see Star Charts on pages 58 and 59.

Chapter 3
An Introduction to Astronomy

The subject of astronomy deals with the Solar System, stars and galaxies, as well as their origin and evolution. The Solar System is made up of our Sun and all the objects that travel about it. The stars are other suns. The galaxies are huge aggregates of hundreds of millions of stars. The universe contains billions (thousands of millions) of galaxies separated by enormous expanses of almost empty space.

The objects which travel around the Sun, and which make up our Solar System, include the eight canonical planets, their one-hundred sixty-six or more moons, tens of thousands of smaller rocky bodies called asteroids, millions of dust-covered icebergs called comets, and a family of icy planetoids in the vicinity of Pluto called Kuiper Belt Objects. All of these reflect light from the Sun. In addition there are an enormous number of small bodies of rock and ice (fragments of asteroids and comets) flying through space. These can be viewed when they pass through the Earth's atmosphere and are heated red-hot by friction, when they are called shooting stars or meteors. Most of them burn up completely. Occasionally one gets all the way through to the surface of the Earth. The rocky or metallic material that survives the descent is called a meteorite.

The stars, like our Sun, are composed of huge and very hot masses of gas. They appear as points of light because they are at enormous distances from us. They are so very far away that looking at them through telescopes does not make them look any larger. The more we magnify the heavens, however, the more stars we can see. What

might seem to be a single dot of light to our naked eye, when magnified many hundreds of times through a large telescope, might appear as a great swarm of suns tremendously far away.

The galaxies appear as fuzzy patches of light with a small telescope. Even a short time-exposed image taken with a moderate-size telescope and a digital detector can reveal the striking spiral or elliptical appearance of these massive collections of stars. Almost all galaxies are so far away that their individual stars cannot be distinguished except with the world's largest telescopes. The spiral galaxy in which our Solar System exists is a fuzzy band of light that stretches across the night-time sky. For this reason, it is called the Milky Way. Even a small telescope (as Galileo discovered in 1610) reveals that the Milky Way is made up of thousands and thousands of stars.

Astronomy changes our ideas of distance, size and time

If we were to set out in an airplane from Honolulu International Airport and fly at 200 miles an hour, it would take us 12 hours to reach San Francisco, 2,400 miles away. If we were to keep on going at that speed, without stopping, we could get all the way around the world in 5 days and 5 hours. If we were able to fly out into space at that speed, it would take us 50 days to reach the moon, and 53 years to fly to the sun. To get to the next nearest star, which happens to be Alpha Centauri, at that speed would take us more than 14 million years!

One of the "yard sticks" which astronomers use to measure the distance to stars is called a "light year." It is the distance that light can travel in one year. Light travels at a speed of more than 186,000 miles a second. To calculate the number of miles in a light year, multiply this by 60x60x24x365.25. Try doing this and see how close the answer comes to six trillion miles. Astronomers have been able to measure the distance to an individual star nearly 900,000 light years away, and measure the distance to a galaxy of stars more than 13 billion light years away.

Stars must be very large to be visible so very far away. We think that the Earth is large (and it is compared to a human), but the diameter of our Sun is more than 100 times the diameter of the Earth. The Sun is just a small star. Some of the stars, whose diameters have been measured, are so large that if we placed the center of one of those stars where our Sun is, the hot gas of that star would reach out beyond the orbit of the Earth and Mars. The distance from the Sun to the Earth is nearly 93,000,000 miles, and from the Sun to Mars 141,600,000 miles. Astronomers and geologists agree that the Earth is more than four and a half billion (4,500,000,000) years old. But even that great length of time is short compared with the overall life span of a star like the Sun. The oldest stars, found in spherical (globular) clusters that orbit the Milky Way and other galaxies, are more than 13.6 billion years old, so the galaxies themselves are probably just a few billions of years older than that.

Apparent movement and location of the heavenly bodies

All the objects in the sky appear to turn about us once each 24 hours. They seem to rise in the east and set in the west. This is caused by the turning of the Earth on its axis from west to east. If you watch the rising or setting of the stars for a few evenings you will see that they appear to have made a small shift westward. A star will appear to rise almost four minutes earlier than it did the evening before. This is caused by the movement of the Earth along its orbit around the Sun. It gets all the way around in one year. The exact advance in time of rising or setting can be calculated by dividing 24 hours by 365.25 days. The nightly change is 3 minutes 56 seconds. In the course of a year we can see a parade of all the stars (except a few around the southern pole of the sky) passing by at the same time of night. These two movements, daily rotation of the Earth on its axis and yearly revolution of the Earth around its orbit, give us our time and our calendar, and make navigation possible. More will be said later about these practical uses of astronomy.

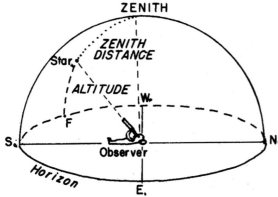

Location of a star by the altitude and azimuth method. The altitude is measured from the star perpendicular to the horizon (F). The astronomer's azimuth would be SF, the angle around the horizon from south to the foot of the perpendicular. The azimuth used by surveyors and navigators would be NF, the angle around the horizon from north through east (E) and south (S) to the foot of the perpendicular from the star.

The celestial sphere

To help locate bodies in the sky, astronomers have devised systems of measurements called coordinates. They know that some stars are comparatively close and that many are very far away, but to make their location simpler they pretend that the stars are all at the same distance. They think of them as being attached to the under side of a great celestial sphere. When we visit a planetarium we see the stars projected on the inside of a great dome. This dome would correspond to half of the celestial sphere. Two different systems of measurements or coordinates are used to locate objects on this sphere.

One system is for the use of an observer at a particular place on the surface of the Earth. This system uses the point directly over the observer's head, called the zenith, and the angle around the horizon, which is everywhere 90 degrees from the zenith. Incidentally, the point on the celestial sphere directly opposite the zenith (underfoot as it were) is called the nadir, but we can't see it, so it does us no practical good. The angle around the horizon, from the north (or south) point, clockwise

through east, south (north), and west, is called the azimuth. Astronomers use the south point, while surveyors and navigators use the north point. With this system, stars are located on the celestial sphere by observing that they are a certain number of degrees (altitude) above the horizon from a point at a certain azimuth. The altitude of the star above the horizon subtracted from 90 degrees gives its angle from the point overhead, which is called the zenith distance. These coordinates are good only for one particular place and time. An observer at any other spot on the Earth's surface would have a different zenith and horizon; a short time later the star would appear to have moved, because of the Earth's turning on its axis, and then it would have a different position as seen by the observer.

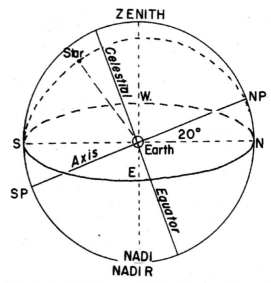

The Celestial Sphere, with the Earth at its center.

The other system of coordinates can be used at any time and place. It is like latitude and longitude on the surface of the Earth. It makes use of the poles of the celestial sphere, which are directly above the north and south poles of the Earth's axis. Half way between these celestial poles is the celestial equator around the celestial sphere. It is everywhere directly above the Earth's equator. On the celestial sphere, declination is measured north and south of this equator, just as latitude is measured north and south of the Earth's equator. Corresponding to longitude on the Earth's surface is right ascension on the celestial sphere. Lines through the celestial poles that cut or intersect the celestial equator at right angles are called hour circles, just as a similar line on the Earth's surface is called a meridian circle. On the Earth's surface, longitude is measured east or west from an observatory near Greenwich, England. On the celestial sphere, right ascension is measured from west to east (increasing in amount as the stars appear to rise) from the vernal equinox or "first point of Aries" (where the Sun appears to cross the celestial equator on its way north, about March 21). The zero hour circle passes through this point of vernal equinox. The right ascension of any star or point on the celestial sphere is the angle, measured along the celestial equator, between the zero meridian and the meridian through the star's point. This may be measured in degrees of arc or as the interval of time between the rising of the one and the other, or of their passing any hour circle across the sky. These circles cross the sky from the northern to the southern celestial pole; the circle through the observer's zenith is his own meridian. Lines of right ascension and declination are marked on the Mercator projection star chart (pages 32 and 33) and illustrate this system of coordinates.

This system of coordinates (right ascension and declination) locates points on the celestial sphere as they might be seen from a point on the Earth's surface which is in line with the object and the center of the Earth. Since the stars are so very far away in comparison with the radius of the Earth (which is only 3,963 miles from the Earth's center to the equator, or 3,950 miles to one of the poles), the coordinates for all practical purposes are good for observers anywhere on the Earth's surface. When making very precise measurements, astronomers have to make small corrections.

Unfortunately this system is not good for all time without correction. The poles of the Earth wobble very slowly, thus changing their position with reference to the stars above them, just as a spinning top wobbles when it begins to run down. One complete Earth wobble takes about 26,000 years. This makes the location of the equinoxes (points where the sun appears to cross the celestial equator) move westward very slowly, about 50 seconds (less than a minute) of arc a year. This is called the precession of the equinoxes. As we noted above, the diameter of the Earth through its equator is 26.7 miles greater than through the poles. The pull of the sun, moon and planets on this slight bulge is what causes the wobbling of the poles. Because of this precession, Vega will become the Pole Star in about 12,000 years, and the point of vernal equinox is now in Pisces.

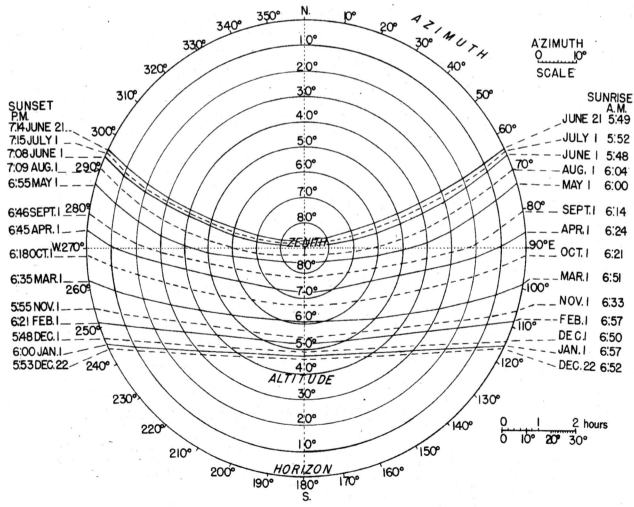

Path of the Sun as seen from Haleakalā, Hawai'i

How the Sun appears to move

The apparent path of the Sun across the sky shifts from south to north and back again in the course of a year. The reason for this is that the Earth's axis is not perpendicular to its orbit around the Sun. During our summer the Earth's northern pole is tipped a little toward the Sun. Six months later, with Earth on the other side of its orbit, it is tipped about the same amount away from the Sun, which shines more directly on the southern hemisphere. The angle of this tip from the perpendicular amounts to about 23.5 degrees. At some time of year the Sun shines directly down on every part of the Earth's surface from 23 ° 27' North Latitude, on June 21 or 22, to 23° 27' South Latitude on December 21 or 22. These parallels mark the limits of the Tropics. The Sun is in the constellation of Cancer when it appears to be furthest north, so this limit is called the Tropic of Cancer. The southern limit is called the Tropic of Capricorn because the Sun is located in Capricornus when furthest south.

The diagram on page 16 shows the way the sun appears to cross the sky as seen from Haleakalā, Hawai'i. The curved lines from side to side represent the tracks along which the Sun appears to move on the first day of each month, and when furthest north (about June 21 or 22), and furthest south (December 21 or 22). The months when the Sun is moving northward (January to June) are solid lines; those when it is moving southward (July to December) are dashed lines.

From this diagram it is possible to tell the direction of the Sun at any hour of any day throughout the year, as seen from Hawai'i. The observer is directly below the dot marked "zenith." The direction of the Sun can be measured as azimuth and altitude. The azimuth is measured around the horizon, starting from the north point; each ten degrees is marked. The altitude of the Sun above the horizon can be estimated from the target of concentric circles, ten degrees apart, from horizon to zenith. Using this diagram, it is possible to estimate the Sun's position to about the nearest degree. This is the angle through which the Sun appears to move in four minutes of time. Time of day is measured from the time of sunrise to that of sunset, along the path lines by means of the little scale in the lower right hand corner, which is two hours of time in length. For example: On August 1, where is the Sun at 5:40 p.m.? This time of day is about an hour and a half before sunset. Measure this amount (three-quarters of the length of the little scale) back along the dash line representing August 1, from the left hand (western) end. This should bring you to the second concentric circle, so you can say at once that the Sun will be 20 degrees above the horizon. A ruler or straight edge through this point

and the "zenith" dot would hit the left hand side of the horizon circle at a point a little less than half way between 280° and 290°. Careful comparison with the "azimuth scale," in the upper right hand corner, would show you that the azimuth is about 284°. This azimuth is the equivalent of 14° north of west or North 76° West. The Sun would set that day at about azimuth 290° at 7:09 p.m.

Another example: When will the Sun be due east? Due east is azimuth 90°. It will rise there on March 22. On April 1 it will be due east about 52 minutes after rising, or at 7:16 a.m., when it is about 12.5 degrees above the horizon. On September 1 it will be due east at 1 hour and 50 minutes after sunrise, or 8:04 a.m., when it is 24 degrees above the horizon. On May 1 it will be due east 3 hours and 27 minutes after rising, or at 9:27 a.m., when it is 48 degrees up the eastern sky, and so on.

When the Sun is furthest north it appears to rise at azimuth 64° at 5:49 a.m.; to cross the meridian at about 3.5° north of the zenith; and to set at azimuth 296° (64 degrees west of north) at 7:14 p.m. When furthest south, the Sun goes almost straight across the sky, from azimuth 116°, when it rises at 6:52 a.m., to azimuth 244°, where it sets at 5:53 p.m. When on the meridian it is about 46.5° above the southern horizon.

This diagram finds practical application in such work as orienting a tennis court so that the players will have a minimum of light in their eyes at any particular time of day. It also can be used to arrange a lānai to get the maximum amount of sunlight in winter and minimum in summer, and for many other architectural and engineering purposes.

Star Chart Finder

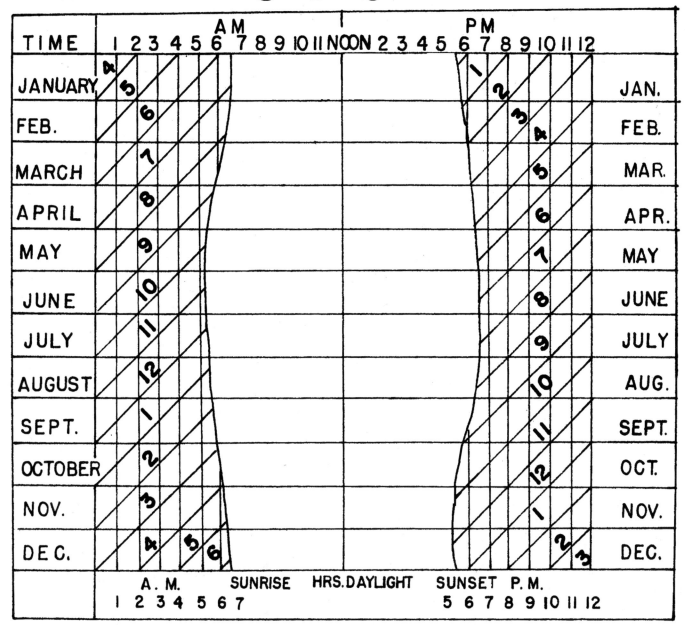

Star charts throughout the night

The monthly star charts are drawn to represent the heavens as seen from the latitude of Hawai'i in the early evening each month, January to December. The diagram above shows how the charts are good also at other hours of the night, and at other times of the year, since the stars appear to rise two hours earlier each month.

Chart 1 is correct on January 1 at about 8 p.m. That same night chart 2 will show the position of the stars at 10 p.m., chart 3 at 12 midnight, chart 4 at 2 a.m., chart 5 at 4 a.m., and chart 6 at 6 a.m. The sun rises at 6:39 a.m. Chart 1 would also represent the heavens at 10 p.m. on December 1; 12 midnight on November 1; 2 a.m. on October 1; and 4 a.m. on September 1.

To choose the correct chart to use: (1) find your date (approximately) by means of the horizontal lines; (2) your time of night by reference to the vertical lines; (3) note where these intersect; and (4) the nearest diagonal line represents the chart number to use, its number standing just above the line. It is possible to interpolate quite closely; for example, at 11 p.m. on June 15, see chart 8 on page 26.

JANUARY
Star Chart 1

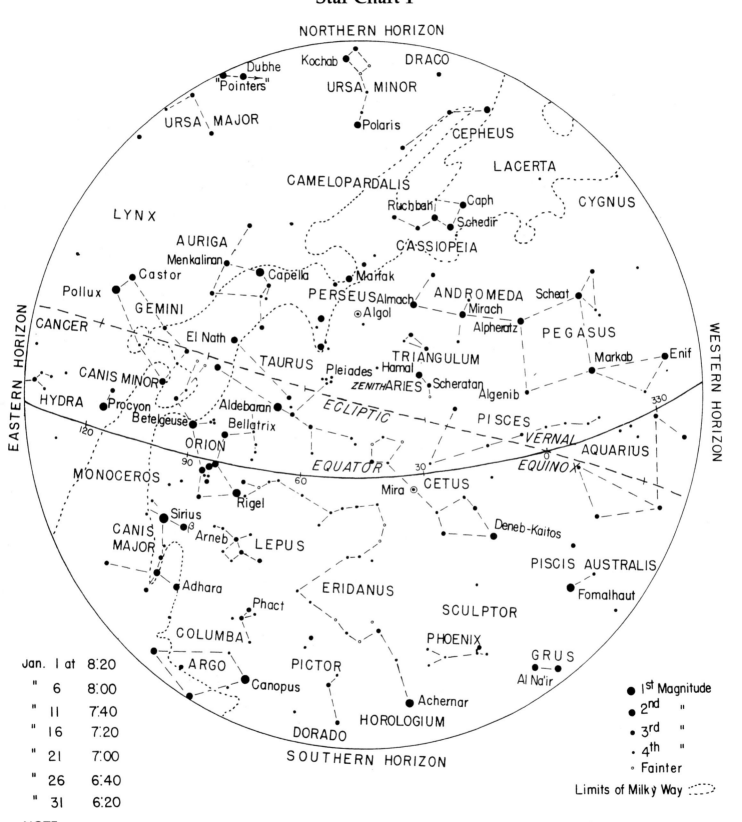

NORTHERN HORIZON

DRACO
Kochab
Dubhe
"Pointers"
URSA MINOR
URSA MAJOR
Polaris
CEPHEUS
LACERTA
CAMELOPARDALIS
CYGNUS
LYNX
Ruchbah
Caph
Schedir
CASSIOPEIA
AURIGA
Menkaliran
Castor
Capella
Marfak
ANDROMEDA
Scheat
Pollux
GEMINI
PERSEUS Almach
Mirach
PEGASUS
CANCER
Algol
Alpheratz
Markab
Enif
El Nath
TRIANGULUM
Hamal
TAURUS
Pleiades
ZENITH ARIES
Scheratan
Algenib
CANIS MINOR
Aldebaran
ECLIPTIC
PISCES
VERNAL
HYDRA
Procyon
Betelgeuse
Bellatrix
AQUARIUS
EQUATOR
30
EQUINOX
ORION
60
CETUS
MONOCEROS
Mira
Rigel
Deneb-Kaitos
CANIS
Sirius
β
MAJOR
Arneb
LEPUS
PISCIS AUSTRALIS
Adhara
ERIDANUS
Phact
SCULPTOR
Fomalhaut
COLUMBA
PHOENIX
GRUS
ARGO
PICTOR
Al Na'ir
Canopus
Achernar
HOROLOGIUM
DORADO
SOUTHERN HORIZON

EASTERN HORIZON

WESTERN HORIZON

Jan.	1 at	8:20
"	6	8:00
"	11	7:40
"	16	7:20
"	21	7:00
"	26	6:40
"	31	6:20

● 1st Magnitude
● 2nd "
• 3rd "
· 4th "
○ Fainter
Limits of Milky Way ⌇⌇⌇

NOTE:
To find the correct chart for viewing times and dates different from those listed please see Star Chart Finder and instructions on page 18. Constellations appear in capital letters. Stars appear in upper and lower case letters.

FEBRUARY
Star Chart 2

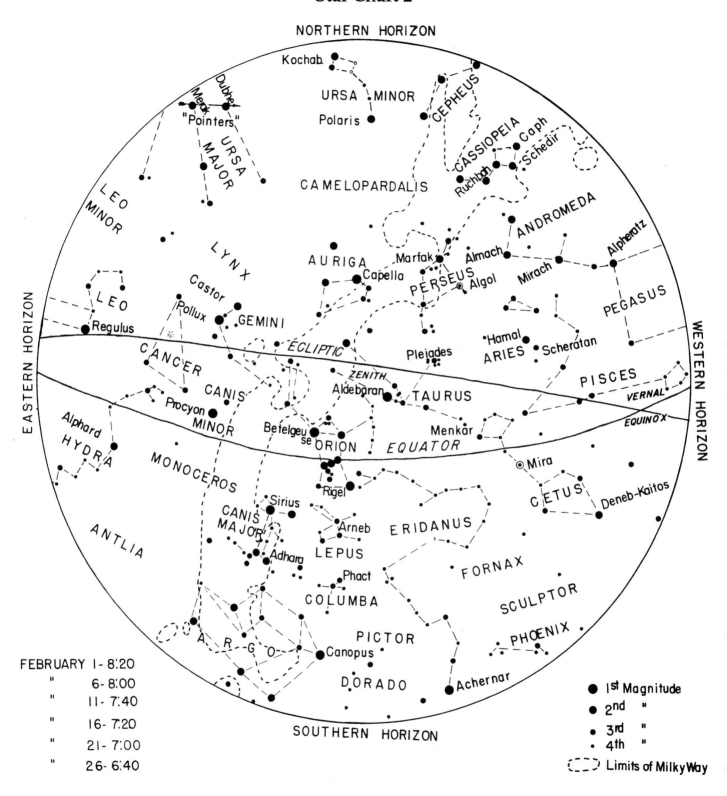

NORTHERN HORIZON

SOUTHERN HORIZON

EASTERN HORIZON

WESTERN HORIZON

FEBRUARY 1- 8:20
"   6- 8:00
"   11- 7:40
"   16- 7:20
"   21- 7:00
"   26- 6:40

● 1st Magnitude
● 2nd   "
• 3rd   "
· 4th   "
⌐⌐⌐ Limits of Milky Way

NOTE:
To find the correct chart for viewing times and dates different from those listed please see Star Chart Finder and instructions on page 18. Constellations appear in capital letters. Stars appear in upper and lower case letters.

MARCH
Star Chart 3

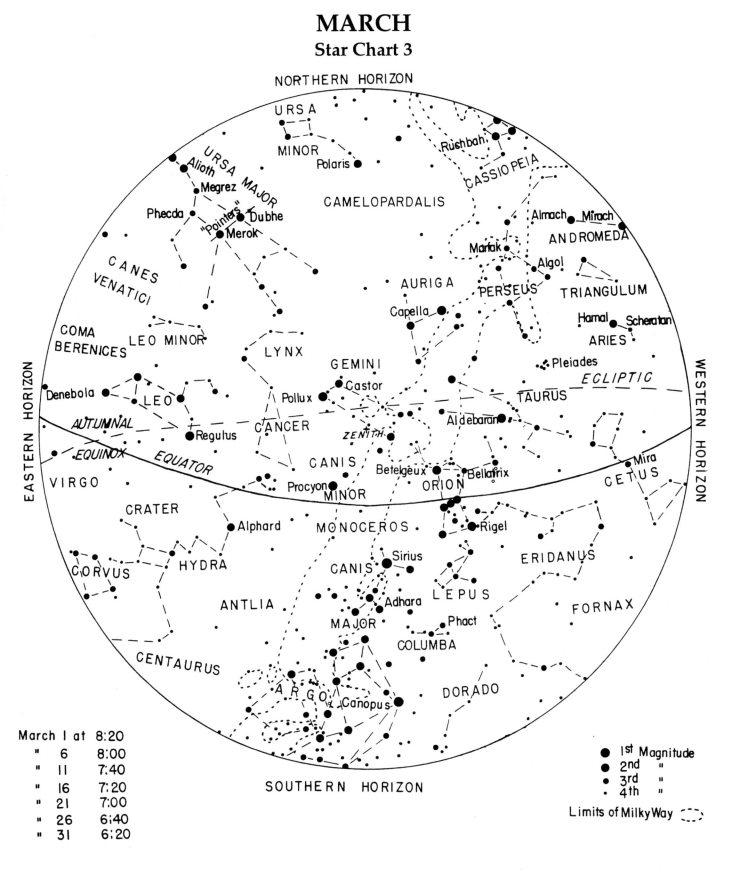

March 1 at 8:20
"   6   8:00
"   11   7:40
"   16   7:20
"   21   7:00
"   26   6:40
"   31   6:20

NOTE:
To find the correct chart for viewing times and dates different from those listed please see Star Chart Finder and instructions on page 18. Constellations appear in capital letters. Stars appear in upper and lower case letters.

APRIL
Star Chart 4

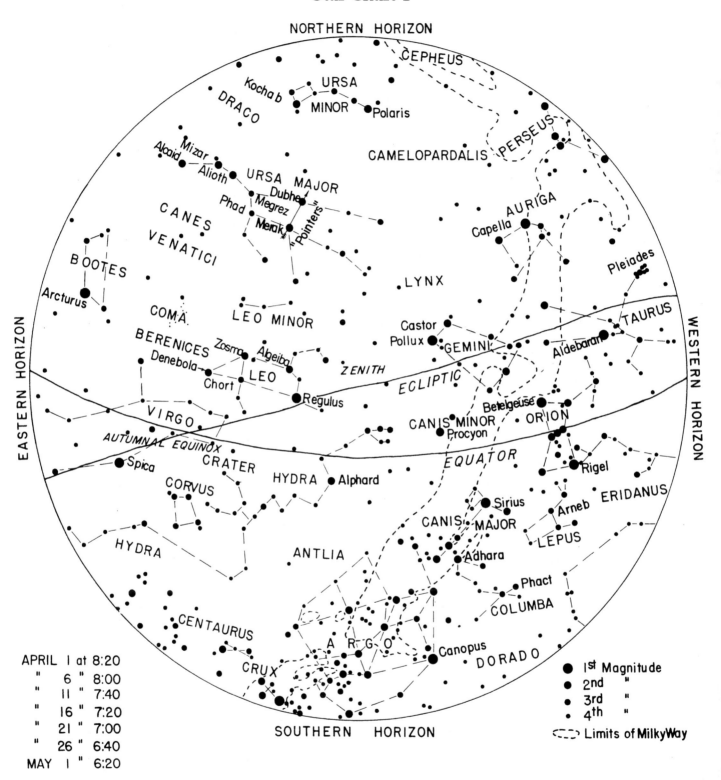

APRIL 1 at 8:20
" 6 " 8:00
" 11 " 7:40
" 16 " 7:20
" 21 " 7:00
" 26 " 6:40
MAY 1 " 6:20

NOTE:
To find the correct chart for viewing times and dates different from those listed please see Star Chart Finder and instructions on page 18. Constellations appear in capital letters. Stars appear in upper and lower case letters.

MAY
Star Chart 5

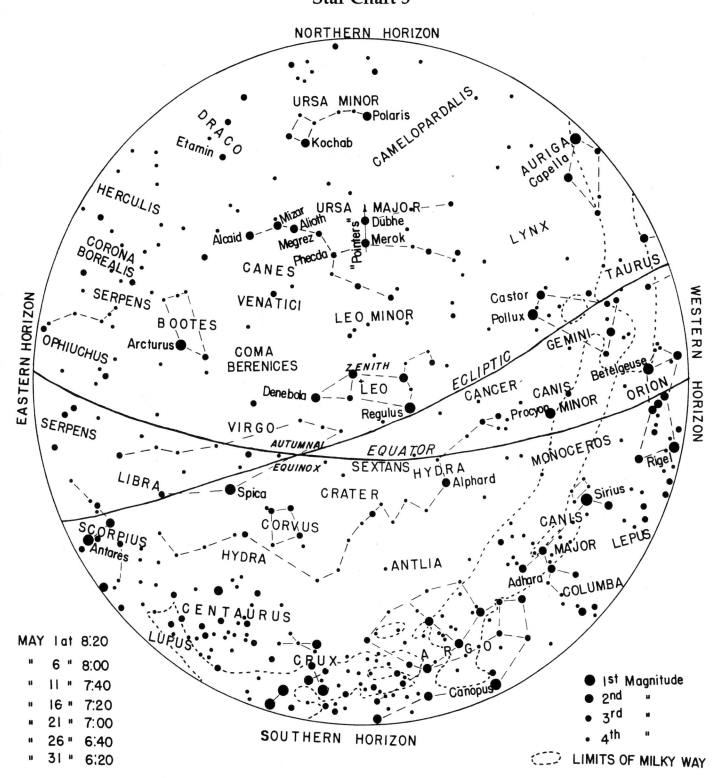

MAY 1 at 8:20
 "   6 " 8:00
 "   11 " 7:40
 "   16 " 7:20
 "   21 " 7:00
 "   26 " 6:40
 "   31 " 6:20

NOTE:
To find the correct chart for viewing times and dates different from those listed please see Star Chart Finder and instructions on page 18. Constellations appear in capital letters. Stars appear in upper and lower case letters.

JUNE
Star Chart 6

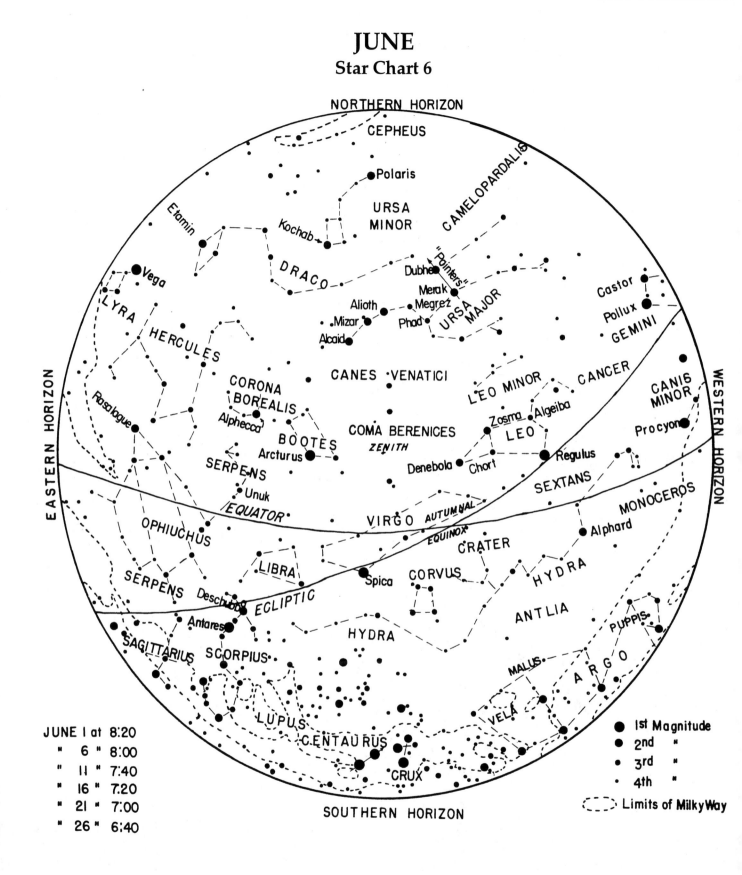

JUNE 1 at 8:20
" 6 " 8:00
" 11 " 7:40
" 16 " 7:20
" 21 " 7:00
" 26 " 6:40

● 1st Magnitude
● 2nd "
• 3rd "
· 4th "
⊂⊃ Limits of MilkyWay

NOTE:
To find the correct chart for viewing times and dates different from those listed please see Star Chart Finder and instructions on page 18. Constellations appear in capital letters. Stars appear in upper and lower case letters.

JULY
Star Chart 7

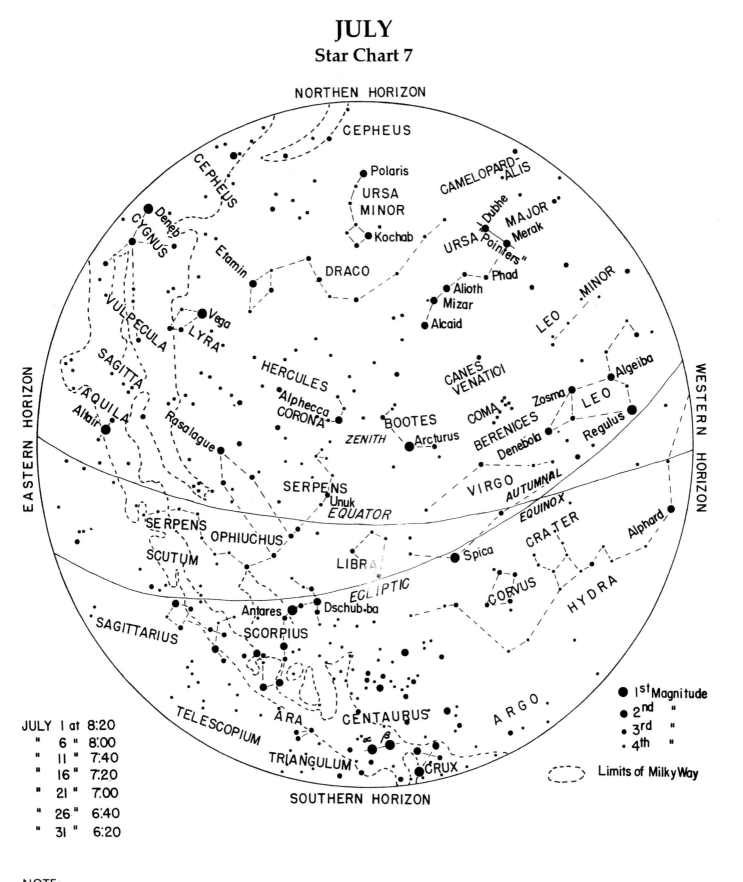

JULY 1 at 8:20
" 6 " 8:00
" 11 " 7:40
" 16 " 7:20
" 21 " 7:00
" 26 " 6:40
" 31 " 6:20

NOTE:
To find the correct chart for viewing times and dates different from those listed please see Star Chart Finder and instructions on page 18. Constellations appear in capital letters. Stars appear in upper and lower case letters.

AUGUST
Star Chart 8

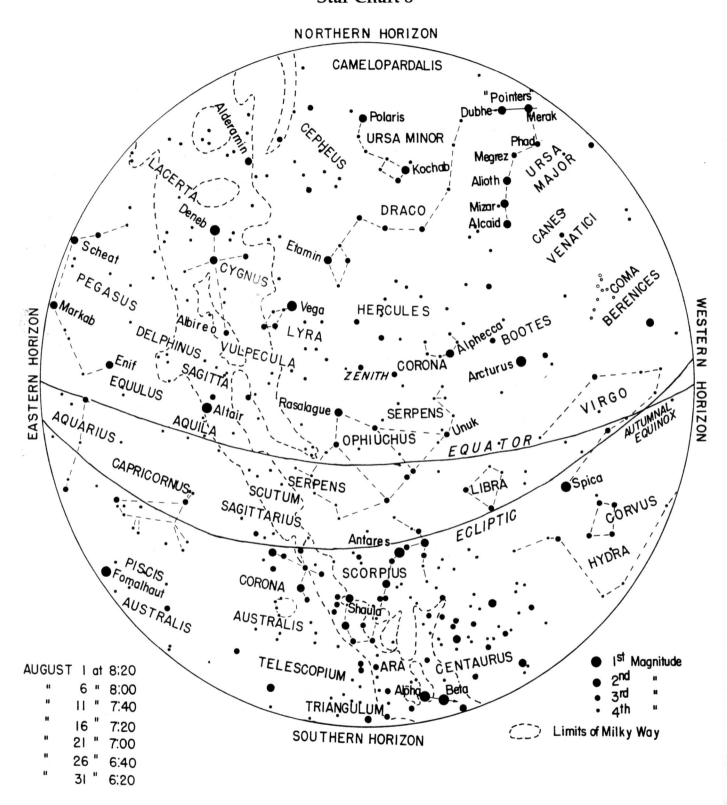

NORTHERN HORIZON

AUGUST 1 at 8:20
  "   6 " 8:00
  "   11 " 7:40
  "   16 " 7:20
  "   21 " 7:00
  "   26 " 6:40
  "   31 " 6:20

● 1st Magnitude
● 2nd "
• 3rd "
· 4th "
⟨⟩ Limits of Milky Way

SOUTHERN HORIZON

NOTE:
To find the correct chart for viewing times and dates different from those listed please see Star Chart Finder and instructions on page 18. Constellations appear in capital letters. Stars appear in upper and lower case letters.

SEPTEMBER
Star Chart 9

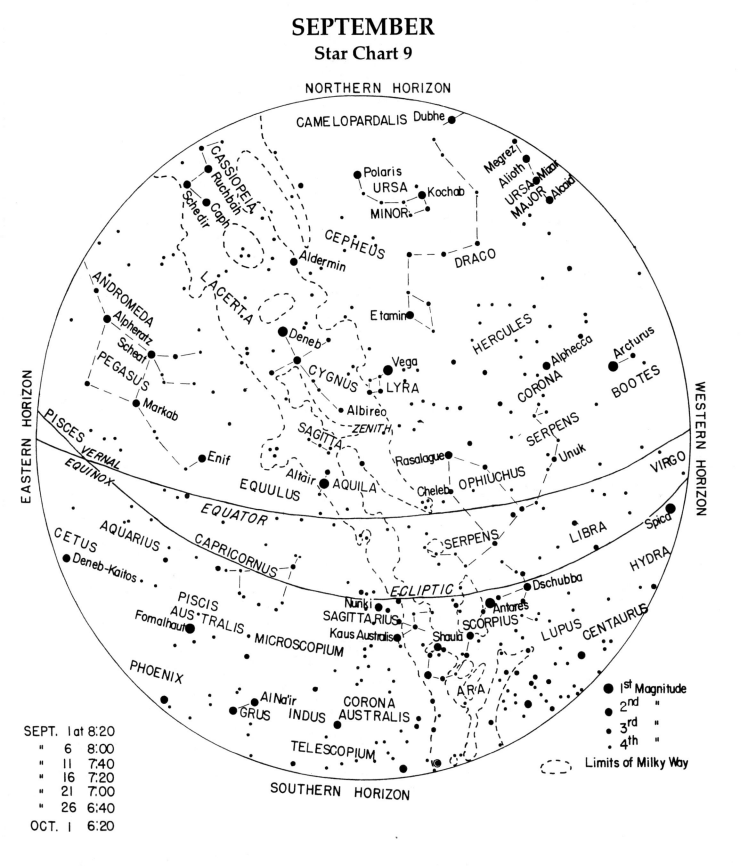

SEPT. 1 at 8:20
"   6   8:00
"   11   7:40
"   16   7:20
"   21   7:00
"   26   6:40
OCT. 1   6:20

NOTE:
To find the correct chart for viewing times and dates different from those listed please see Star Chart Finder and instructions on page 18. Constellations appear in capital letters. Stars appear in upper and lower case letters.

OCTOBER
Star Chart 10

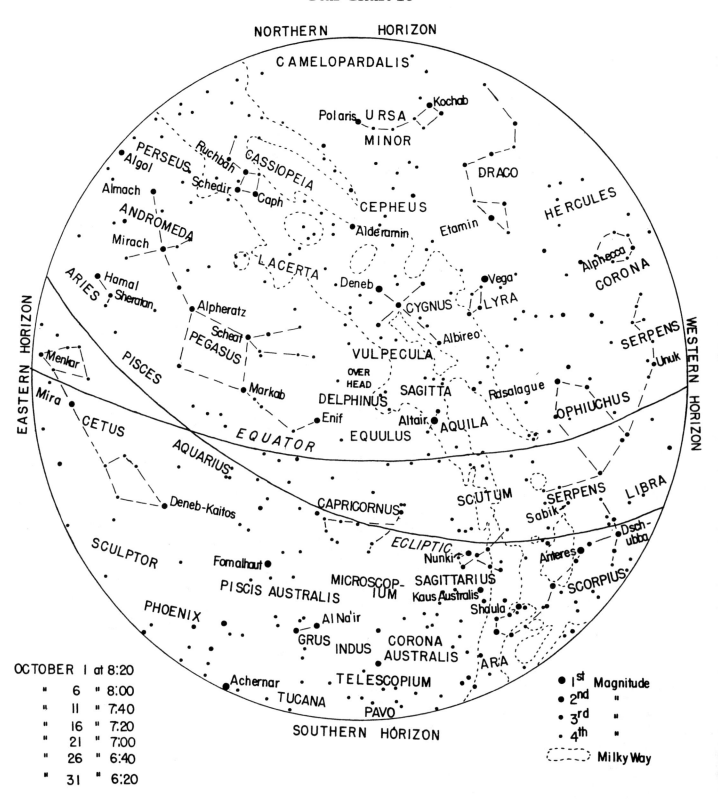

OCTOBER 1 at 8:20
"   6   " 8:00
"   11   " 7:40
"   16   " 7:20
"   21   " 7:00
"   26   " 6:40
"   31   " 6:20

NOTE:
To find the correct chart for viewing times and dates different from those listed please see Star Chart Finder and instructions on page 18. Constellations appear in capital letters. Stars appear in upper and lower case letters.

NOVEMBER
Star Chart 11

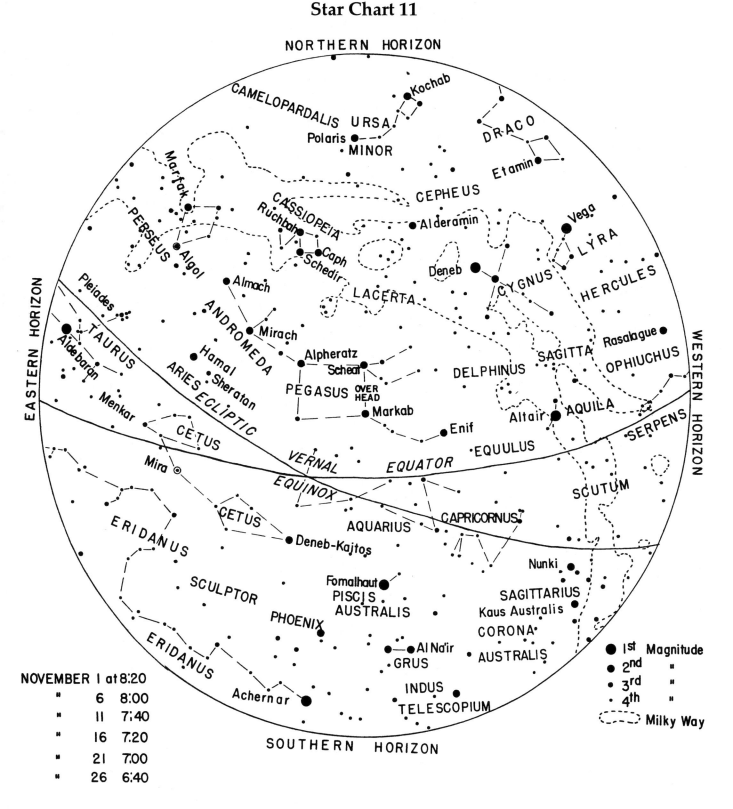

NOVEMBER 1 at 8:20
"   6   8:00
"   11   7:40
"   16   7:20
"   21   7:00
"   26   6:40

NOTE:
To find the correct chart for viewing times and dates different from those listed please see Star Chart Finder and instructions on page 18. Constellations appear in capital letters. Stars appear in upper and lower case letters.

DECEMBER
Star Chart 12

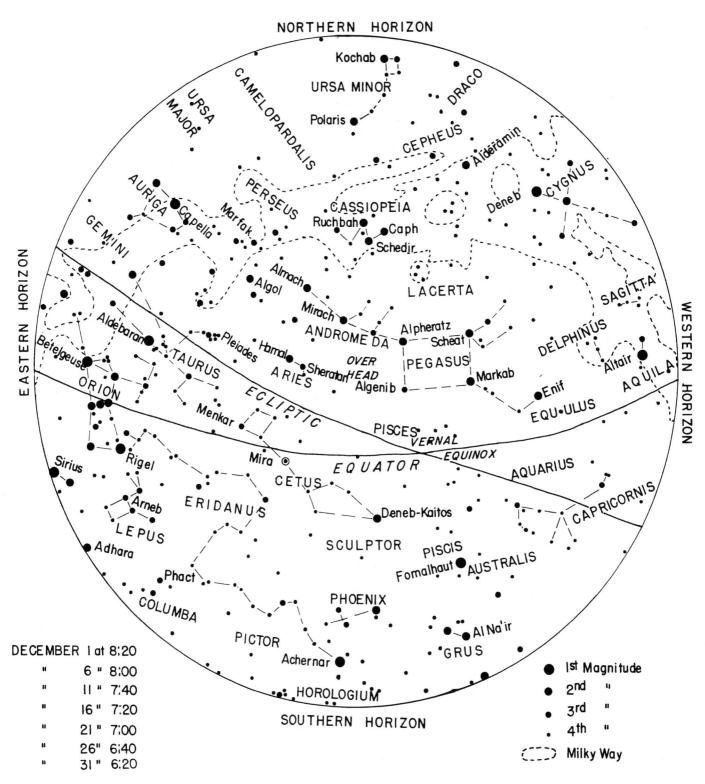

DECEMBER 1 at 8:20
  "   6 " 8:00
  "   11 " 7:40
  "   16 " 7:20
  "   21 " 7:00
  "   26 " 6:40
  "   31 " 6:20

NOTE:
To find the correct chart for viewing times and dates different from those listed please see Star Chart Finder and instructions on page 18. Constellations appear in capital letters. Stars appear in upper and lower case letters.

NORTHERN SKY
North Polar Stars
Star Chart N

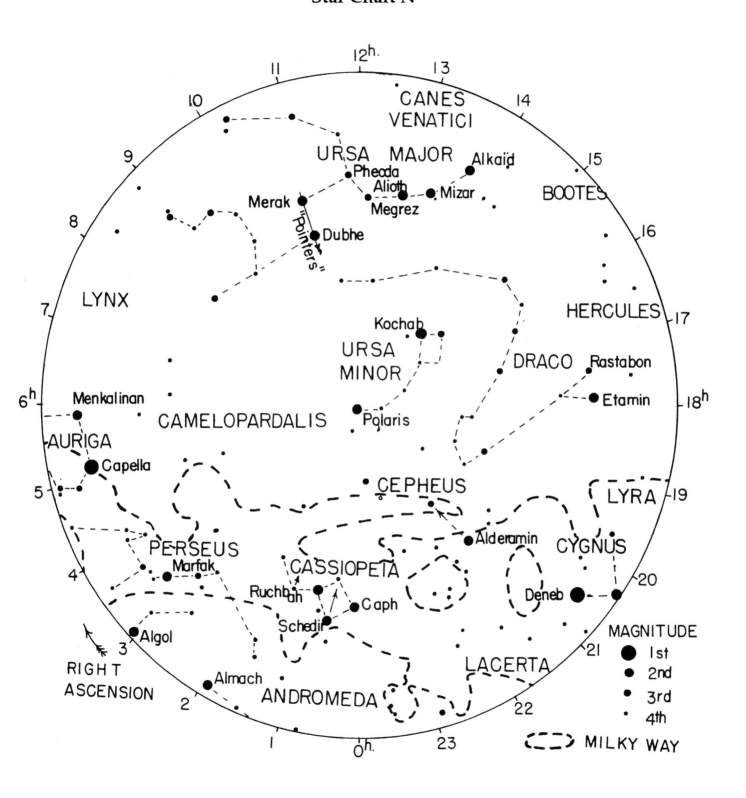

URSA   MINOR

340°   320°   300°   280°   260°   240°   220°   200°   18

80°

Kochab

CEPHEUS

DRACO

60° Alderamin

Megrez
Alioth
Mizar
Alkaid   URSA

Etamin

Deneb

Vega

HERCULES

Cor
Caroli

40°

CYGNUS

LYRA

CORONA
BOREALIS   CANES VEN

Albireo

Alphecca   BOOTES

Scheat

Arcturus

PEGASUS

VULPECULA

20°

SAGITTA
DELPHINUS

SERPENS

Denebol

Markab

Rasalague

Enif
EQUUL-
US   AQUILA

Altair

OPHIUCHUS   Unuk

**VERNAL
EQUINOX**

EQUATOR

AQUARIUS

SCUTUM

VIRGO
Spica

LIBRA

CAPRICORNUS   ECLIPTIC

CORVUS

-20°

PISCIS

Nunki

Antares

Fomalhaut

SCORPIUS

AUSTRALIS

SAGITTARIUS

Kaus Australis

CENT

-40°

MICRO-
SCOPIUM

CORONA
AUSTRALIS

LUPUS

CRUX

GRUS

Al Na'ir

NORMA

PAVO

ARA

β

-60° TUCANA

TELESCOPIUM

CIRCINUS   α   β

β

Rigil Kentaurus

α

INDUS

23   22   21   20   19   18

MUSC

TRIANGULUM
AUSTRALIS

-80°

17   16   15   14   APUS   CH

13   12

● 1ˢᵗ Magnitude   • 3ʳᵈ Magnitude
● 2ⁿᵈ   "   · 4ᵗʰ   "   (in part, to fill out figures)

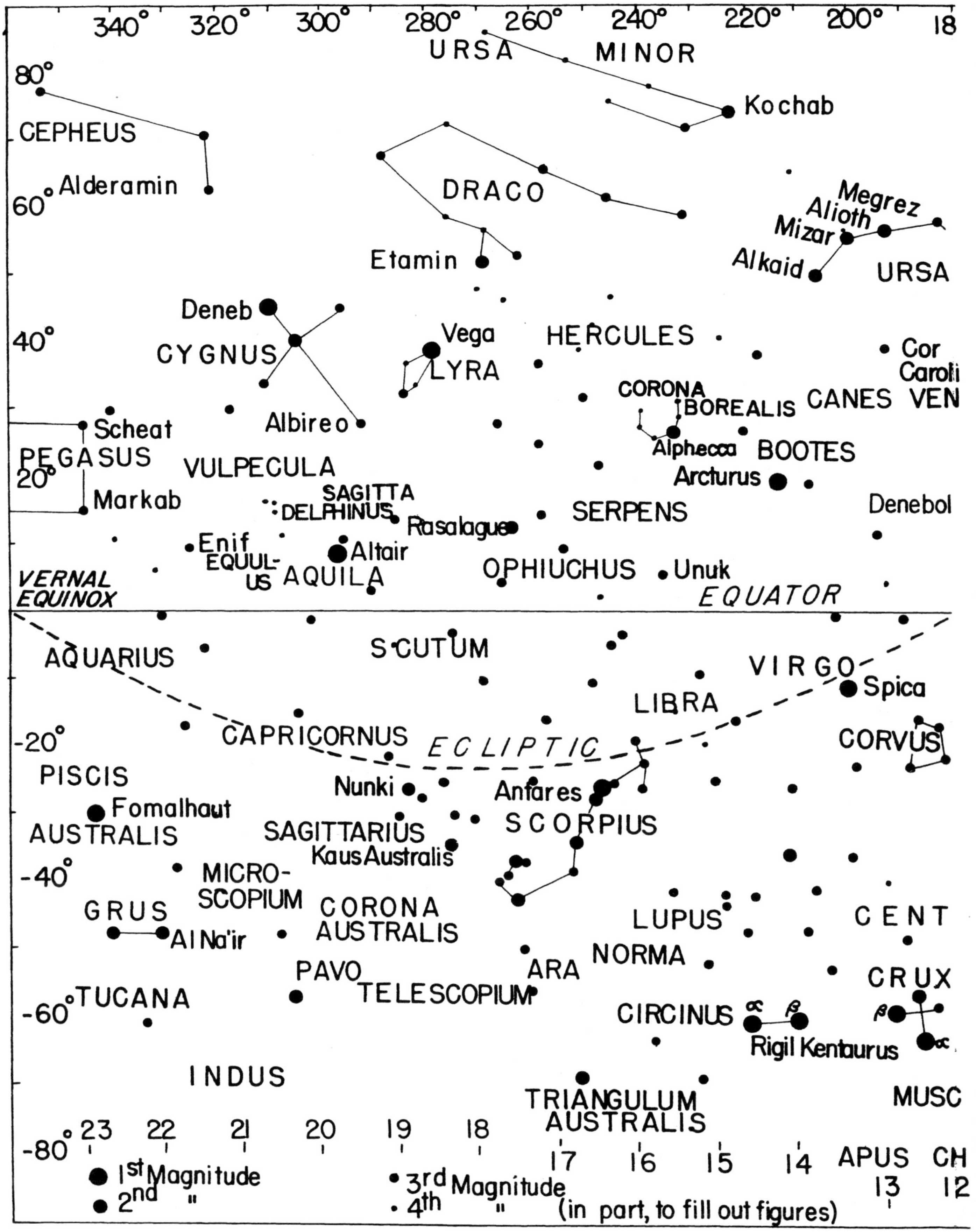

The "Mercator" Star Chart shows all of the constellations and brighter stars around the whole sky, but not as they would look

STAR CHART

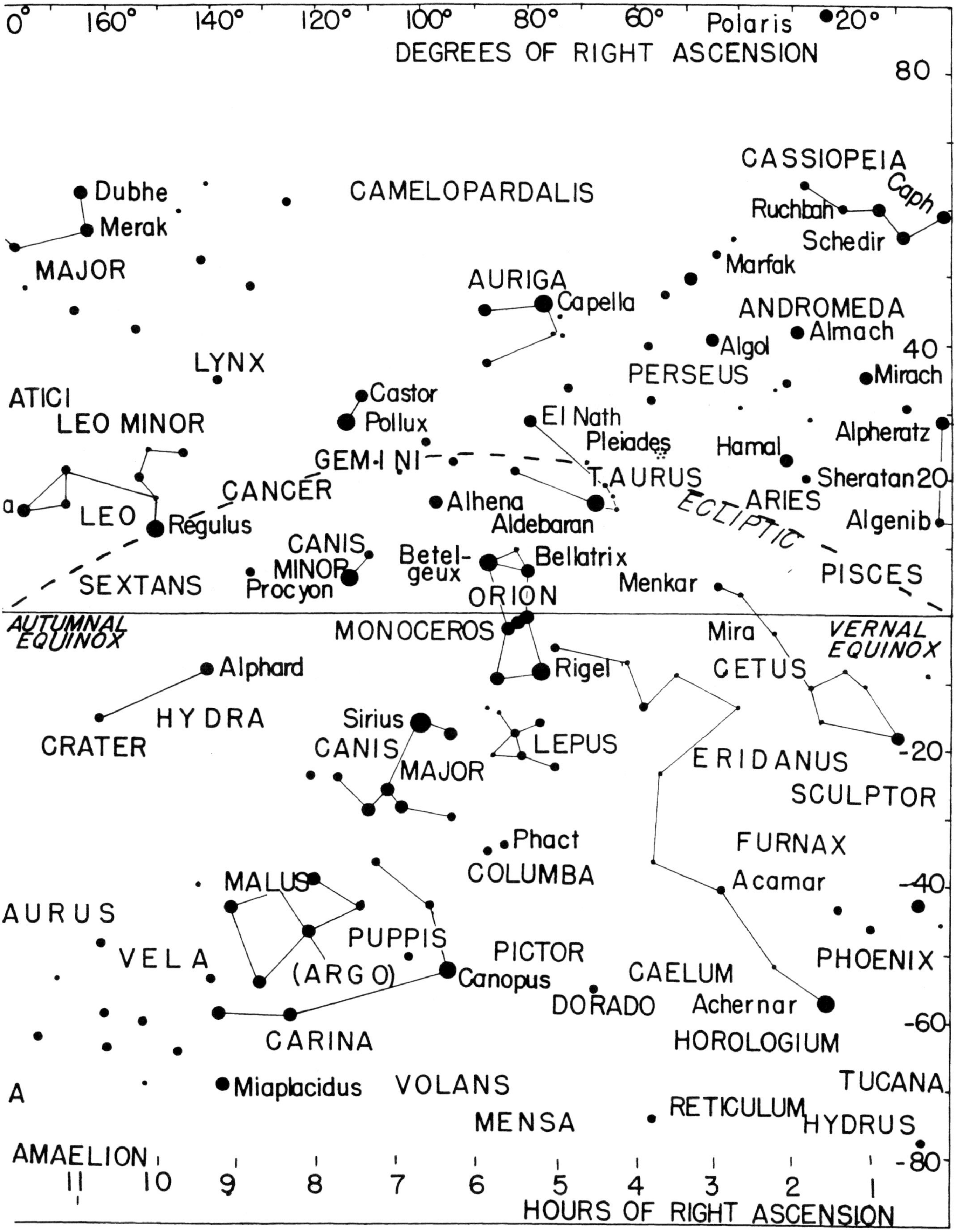

DEGREES OF RIGHT ASCENSION

0° 160° 140° 120° 100° 80° 60° Polaris 20°

80

CASSIOPEIA
Caph
Ruchbah
Schedir
Marfak

Dubhe
Merak
MAJOR

CAMELOPARDALIS

ANDROMEDA

AURIGA
Capella

Algol
Almach 40
Mirach

LYNX

PERSEUS

ATICI
LEO MINOR

Castor
Pollux
GEMINI

El Nath
Pleiades

Hamal

Alpheratz

Sheratan 20

ARIES

CANCER

Alhena

TAURUS

ECLIPTIC

a

LEO Regulus

Aldebaran

Algenib

CANIS
MINOR
Procyon

Betel-
geux

Bellatrix

Menkar

PISCES

SEXTANS

ORION

AUTUMNAL
EQUINOX

MONOCEROS

Mira

VERNAL
EQUINOX

Alphard

Rigel

CETUS

HYDRA

CRATER

Sirius

CANIS

MAJOR

LEPUS

ERIDANUS

SCULPTOR

Phact

COLUMBA

FURNAX

Acamar

AURUS

MALUS

VELA

PUPPIS
(ARGO)

PICTOR

Canopus

CAELUM

DORADO

Achernar

PHOENIX

CARINA

HOROLOGIUM

A

Miaplacidus

VOLANS

MENSA

RETICULUM

TUCANA

HYDRUS

AMAELION

11 10 9 8 7 6 5 4 3 2 1

HOURS OF RIGHT ASCENSION

-20

-40

-60

-80

at any particular time from Hawai'i. This chart illustrates the use of right ascension and declination to locate stars in the sky.

SOUTHERN SKY
South Polar Stars
Star Chart X

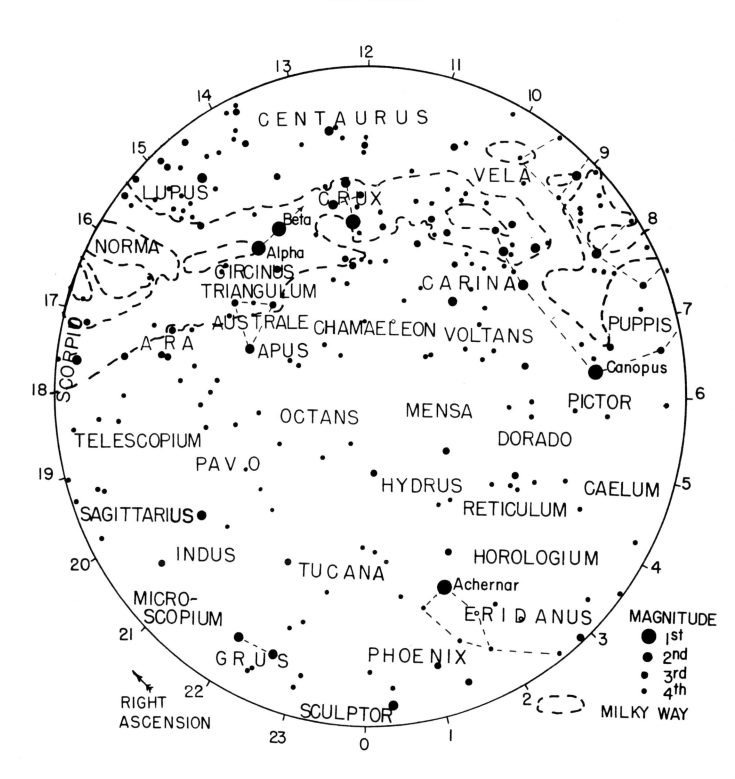

Chapter 4
The Solar System

As noted in Chapter 3, the Solar System is made up of the Sun, its eight known planets, their moons, numerous asteroids and Kuiper Belt Objects, a number of comets, and great quantities of smaller pieces of meteoric material that are seen when they pass through the Earth's atmosphere and are heated red hot by friction.

The Sun, our nearest star, is the largest and most important member of the Solar System. Without the light, heat, and energy that it furnishes, no life (as we know it) would be possible on the Earth or on the other planets.

The Sun is a huge ball of exceedingly hot gas under enormous pressure. Its mean diameter is about 864,100 miles, and its mean distance from the Earth is 92,960,000 miles. The sun rotates on its axis at varying rates (from 25 to 36 days for one rotation, fastest at the equator). The temperature at the Sun's surface is about 5,800 degrees absolute (9,980 degrees Fahrenheit). The temperature increases toward the Sun's center, where it exceeds 10 million degrees absolute. Names have been given to the outer layers of gas (all that are visible) near the Sun's surface. From the deepest visible layer (less than 300 miles down) of the Sun's atmosphere outward, we see: the photosphere ("light-giving stratum"), where some of the light from deeper strata is absorbed, leaving dark lines in the Sun's spectrum; the chromosphere, a low-density layer through which tongues and jets of hot gas are continually moving; and the corona, an extended irregular halo of very tenuous gas, much hotter than the Sun's surface, that is best seen during a solar eclipse. The jets of chromospheric gas, called solar prominences, shoot many thousands of miles out from the Sun's surface. Dark spots (sunspots), some of large size (much larger than the Earth) can be seen on the Sun's surface. These increase in number and change their position, moving toward the Sun's equator, during a period of a little over 11 years. There is also a cycle of 22-year magnetic activity on the Sun, which correlates with the sunspot number; thus, sunspots have been found to be active magnetic regions at the base of the photosphere which extend up through the chromosphere to the corona.

Each square meter of the Sun's surface radiates energy equal to 64,000,000 Watts (85,000 horsepower) continuously. Such vast amounts of energy are released by nuclear fusion at the Sun's core, where the temperature is 15 million degrees absolute. At these high temperatures, four hydrogen nuclei (protons) fuse into a single helium nucleus. The helium nucleus is 0.7% less massive than the total mass of the four protons, and the difference is transformed into light energy. By this process, the Sun is converting 660 million tons of hydrogen gas into helium gas every second; it is actually radiating away 4 million tons each second, yet because the Sun has such an enormous mass (333,000 times that of Earth), fusion of hydrogen into helium (called hydrogen "burning") can continue for about 10 billion years for a star of the Sun's mass. Curiously, more massive stars have shorter hydrogen-fusion lifetimes; their core temperatures are much higher, so the nuclear fusion uses up the hydrogen much faster. They are like brighter "candles" that "burn out" faster. A star 10 times more massive than the Sun uses up its hydrogen fuel 100 times faster, which corresponds to a hydrogen-fusion lifetime of 100 million years; a star 50 times more massive than the Sun uses up its hydrogen fuel in only 4 million years. There are small changes in the Sun's heat and light output, but there has not been any overall marked decrease during geologic time, so we can expect the Sun to keep on giving heat and light to its planets for many billions of years to come. However, when all the hydrogen gas in the Sun's core is converted into helium gas, a drastic change in its structure will take place. In about 5 billion years, the Sun will bloat up, becoming a red giant (like Aldebaran or Arcturus), and will engulf Mercury and Venus; here on Earth, the oceans will vaporize and all life will likely be extinguished.

RELATIVE DISTANCE OF THE PLANETS FROM THE SUN

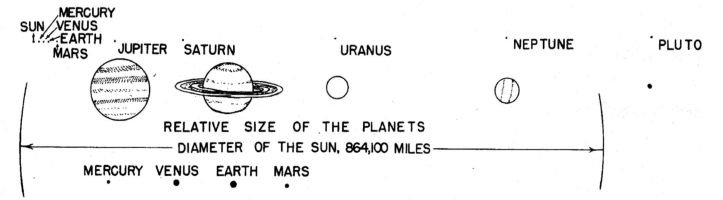

RELATIVE SIZE OF THE PLANETS

DIAMETER OF THE SUN, 864,100 MILES

MERCURY   VENUS   EARTH   MARS

The Planets and a Brief History of Astronomy

Some of the more important facts about the planets are given in the tabulation on page 43. Mercury, Venus, Mars, Jupiter, Saturn, and the Earth's moon have been known from antiquity. William Herschel discovered Uranus in his reflecting telescope, March 13, 1781. The planets pull on each other and this keeps their orbits from being exact circles. The irregular, elliptical orbit of Uranus led two persons to predict independently that there was another planet out beyond. These predictions were made by John C. Adams (October 1845) and Urbain Leverriere (September 1846). As a result, Neptune was discovered at the Berlin Observatory, September 23, 1846. Similar predictions and calculations (shown later to be erroneous) were made about another planet beyond Neptune, and it was discovered at Lowell Observatory, March 13, 1930, by Clyde Tombaugh, and subsequently named Pluto. However, Pluto's mass is far too small to significantly perturb the orbits of either Uranus or Neptune; thus, the discovery of Pluto was a lucky accident driven by miscalculation and faith.

Some of the ancient philosophers believed the Earth to be round, to rotate on its axis, and like other planets, to revolve around the Sun. However, most believed that the Earth must be the center of everything, and that therefore the Sun and the planets, and also the stars must go around the Earth. The Greek mathematician Ptolemy (ca. 140 C.E.) and his followers used a system of epicycles to explain the movements of the planets, which they believed traveled in circular paths around the Earth. In Ptolemy's somewhat complex model, each planet moved around a point (deferent) that went around the Earth. This scheme successfully accounted for the observed motions for about 1400 years, but eventually errors in the calendar based on these apparent motions became noticeable, and Nicolaus Copernicus (1473-1543) was asked to examine the problem. The result of this was that Copernicus formulated a new model in which all of the observed movements could be explained simply if the Sun were the center and the Earth moved around it in a circular path between the paths of Venus and Mars. Copernicus published this work on his deathbed in 1543, but the Catholic Church was initially not alarmed about the new model since it was seen as a convenient mathematical fiction. When Galileo Galilei (1564-1642) publicly supported the Sun-centered model in print, and portrayed believers in the Earth-centered model as simpletons (ca. 1616), he was forced by the Church to deny these views under oath. Galileo was the first to use the telescope to show by observation that the Sun-centered model is correct: Venus goes through a complete cycle of phases, showing that it must move around the Sun and not the Earth (thus refuting Ptolemy's model), and Jupiter has four large moons that move around it, showing that Earth is not the center of everything in the Solar System.

Johannes Kepler (1571-1630) became a "believer" in the Copernican system early in his career. He inherited a wealth of high-quality observational data collected by the Danish astronomer Tycho Brahe (1546-1601), and discovered that the planets travel in elliptical orbits, such that they move more quickly when closer to the Sun and move more slowly when farther away from the Sun. He also discovered a strict mathematical relationship between the distance of a planet from the Sun and the time taken to complete one revolution relative to the stars. This showed that a physical force was causing planetary motions. That physical force was identified by Isaac Newton (1642-1727) as gravity. Newton showed that the force of gravity exerted by the Sun on Earth and the other planets keeps them travelling in curved paths (orbits); without that force, the planets would move off in straight-line paths. Since the force of gravity weakens with distance, planets (of similar mass) which are farther away move more slowly. Newton was thus able to derive Kepler's mathematical relationship from his principle of universal gravitation.

The Earth's Moon

The most important and conspicuous moon is the single one that circles about the Earth at a mean distance of 238,860 miles. Because its orbit is not a perfect circle, it can get as close as 222,000 miles and as far away as 253,000 miles. Its diameter is about 2,160 miles, but it isn't a perfect sphere. It is about three-fifths as dense as the Earth (3.3 as compared with the Earth's 5.5 times the density of water). Thus it weighs only one and a quarter percent (0.0123) as much as the Earth does. The full moon gives as much light as a quarter of a million candles; but it would take the light from over 600,000 full moons to equal sunlight. The moon always keeps the same face toward the Earth, as if it were on the end of a string, this string being the force of gravitation. It takes 27 days, 7 hours, 43 minutes and 11.5 seconds for the moon to make one circuit of the heavens (a sidereal month); but from new moon to new moon (synodic month) is 29 days, 12 hours, 44 minutes and 2.8 seconds, a headache for the calendar makers.

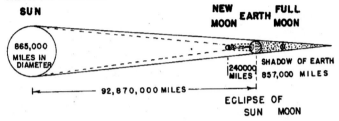

Two interesting features of the moon are eclipses and the tides that it helps to cause on the Earth. An eclipse of the moon is caused when the moon (at full phase) passes, either partially or entirely, through the Earth's shadow. An eclipse of the Sun occurs when the moon (at new phase) gets between the Earth and the Sun, such that its shadow intersects the Earth's surface. A total eclipse does not occur at every new or full moon because the plane of the moon's orbit is tipped 5 degrees with respect to the plane of the Earth's orbit. Thus, we can have eclipses only when the line of intersection between these two planes is aligned with the Sun-Moon-Earth or Sun-Earth-Moon directions. This alignment occurs every 173.3 days, a little less than six months apart. Figure above shows how these configurations occur. Lunar eclipses are about as frequent over the entire Earth as solar eclipses when we count partials of both types. However, from a given place on Earth, one is more likely to see a total lunar eclipse than a total solar eclipse, because everyone on the night side of the Earth is in the Earth's shadow, while the Moon's shadow at the distance of the Earth is only 170 miles wide at maximum. There are from two to seven (most of them partial) eclipses each year. The minimum number is two of the Sun; the maximum, two of the moon and five of the Sun, or three of the moon and four of the Sun. From 1901 to 2000 C.E., there were 375 eclipses, 228 of the Sun and 147 of the moon. Eclipses run in cycles, the exact pattern recurring after 6,585.32 days (18 years, 10.7 days). This cycle is called a saros. However, since the interval is not an integral number of days, the next eclipse in the cycle occurs one-third of the way around the Earth from the previous one, so we must wait 3 saros cycles (54 years, 32 days) for the next total solar eclipse from the same place on Earth. Total solar eclipses occur every 1.5 years on the average over the Earth, but from one place, the mean frequency is 1 every 360 years.

Tides are produced by the pull of the Sun and the Moon on the surface of the earth. We don't notice Earth tides, but the pull of these heavenly bodies is able to make an appreciable bulge on the surface of water, especially noticeable where this runs up onto land as into a bay. Tides are caused by differences in gravitational pull across an object by a second object; the side of the Earth closest to the Moon feels a greater pull than the Earth's center does, separating the near side from Earth's center; similarly, the Earth's center feels a greater pull than the far side of the Earth does, separating Earth's center from the far side. The Earth and Moon, due to mutual tidal forces, become football-shaped bodies (oblate spheroids). Because of the rigidity of the Earth, its solid mass does not reach an equilibrium shape. Since the ocean water flows freely over the surface, the water flows horizontally over the Earth so as to pile up at the point closest to the Moon and to the point on the opposite side of the Earth from the Moon. Two tidal bulges thus form, and as the Earth rotates, each point on the surface experiences two high tides and two low tides a day. Landlocked bodies of water do not experience significant tides. The Sun's tidal force on the Earth is about half as great as that of the Moon, since the Moon is so much closer, and thus the Moon's differential gravitational force across the Earth is larger.

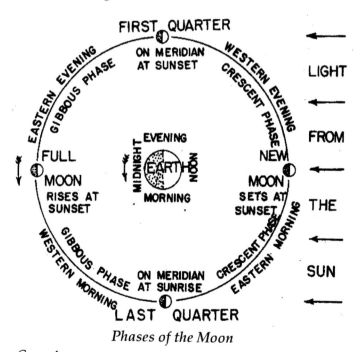

Phases of the Moon

Comets

Comets are comparatively small, dust-covered chunks of ice which move around the Sun in highly elliptical orbits. As a comet gets close to the Sun, it develops a long tail which is always aimed away from the Sun, regardless of the direction in which the comet itself is moving. The mass of a comet is very small, most of this being in its "nucleus," which may be from 10 to 15 miles in diameter. The nucleus can be described as a "dirty iceberg." Bits and pieces of dust and rocky material are mixed in with water-ice. The tail, if and when it becomes visible, may be as much as a hundred million miles in length. The gas tail consists of streams of radiating charged molecules driven away from the nucleus by collisions with the solar wind particles, while the dust tail consists of larger grains pushed away by the Sun's rays which reflect sunlight. Perhaps as many as 1,000 comets a century circle the Sun, but less than one in five becomes visible to the naked eye, and only a dozen a century become conspicuous. Only about a third of those that go out into space return again. Halley's comet is one of the most famous. Its return, at intervals of about 76 years, has been observed for centuries. The last approach to the Sun was in 1986; the next should occur in 2062.

Name of Meteor Shower	Duration (days)	Date of Maximum	Meteors per hour
Quadrantids	15	January 4	120
Lyrids	9	April 22	18
Eta Aquarids	39	May 5	65
Arietids	45	June 7	54
June Bootids	15	June 16	18
Perseids	36	August 12	100
Orionids	35	October 21	25
Taurids	52	November 5-6	10
Leonids	24	November 17	15
Geminids	10	December 13	120
Ursids	9	December 23	10

Meteors

Popularly called shooting stars, these are small to medium size pieces of rock and ice. Many have been analyzed, after having passed all the way through the Earth's atmosphere without burning up. Such pieces are known as meteorites. Swarms of meteors travel in orbits around the Sun and many millions of them are believed

to enter the Earth's atmosphere every day. The orbits of these swarms have been associated with those of former comets. In parts of these orbits they are more numerous than in others, so that at certain times each year the Earth passes through these swarms and meteors become very numerous. The preceding table (from the American Meteor Society) notes the names given to these recurring meteor showers, their duration, and the dates when they usually reach their maximum display. The names usually refer to the constellation from which they appear to radiate.

Asteroids

The large gap in spacing between Mars and Jupiter led astronomers of the 18th century to speculate that there was a planet at that location. After a careful search, a small object (subsequently named Ceres) was discovered in 1801; its diameter was estimated to be only 600 miles. Shortly after, three smaller objects were found (named Pallas,

Juno and Vesta). After photographic techniques were developed, many more of these asteroids were discovered. It is estimated that about 100,000 of these irregularly-shaped objects exist. Ceres is the largest, but the total mass of all the asteroids amounts to 0.1% of the mass of the Earth.

The Earth has been impacted by asteroidal objects in the past. The mile-wide Barringer Crater in Arizona - most commonly called Meteor Crater - is only about 50,000 years old. Created by a 160 foot wide nickel-iron meteorite traveling about 28,600 mph that mostly vaporized on impact, the crater is one of the Earth's best preserved impact craters. Another famous example is the June 30, 1908, explosion which occurred over a remote region of Siberia. This "Tunguska Event" released the energy of a thermonuclear blast, but without the radioactivity. A forest 20 miles in diameter was completely burned, and trees fell like "matchsticks" away from a central location where no crater or large meteoritic fragment was ever located. Recent calculations suggest that the Tunguska event was likely caused by the breakup of a stony asteroidal fragment 25 feet across in the low atmosphere. It is now believed by many astronomers that the extinction of the dinosaurs 65 million years ago can best be explained by the impact of a 6-mile asteroid on the Yucatan peninsula of Mexico. Evidence for this includes a shallow crater 100 miles wide off the coast near the village of Chicxulub, an iridium-rich deposit of dark clay sandwiched between 65-million-year-old limestone layers, and impact-shocked quartz crystals of that same age discovered at the Chicxulub site.

How the Solar System was formed

After the Copernican Sun-centered model had become universally accepted, scholars began to wonder how such a system had been formed. Philosophers, such as Immanuel Kant and Emanuel Swedenborg, proposed elaborate theories about the origin of the Solar System. These men were neither mathematicians nor astronomers, but their ideas were taken and developed by a French astronomer, Pierre Simon Marquis de Laplace. His Nebular Hypothesis suggested that the Sun and all the planets and their moons were once a huge nebula or circular, flat mass of hot gas, which extended out beyond the orbit of the outermost planet, and which rotated. As this mass of gas cooled it contracted and rotated faster. This gave the gas on the outer edge inertial motion away from the center. When this "centrifugal" force balanced the pull of gravity toward the center, a ring of gas was left behind as the rest of the nebula continued to contract. The revolving ring of gas eventually condensed into a globular mass, which contracted and formed a planet rotating on its axis. Other rings, left behind as the nebula continued to contract, formed the other planets in a similar manner. In some cases small rings formed around the gaseous planets, as they contracted, and became their moons. The ring between those that formed Jupiter and Mars did not condense into a planet, but instead formed thousands of asteroids (these were not discovered until after 1801, but were incorporated into the Nebular Hypothesis at that time). The material left in the center became the present Sun. Observed points in favor of this theory were the rings around Saturn and the numerous flat, circular or spiral nebulae, then recently discovered and believed to be composed of gas. These "spiral nebulae" used to be considered solar systems in the making. Today, with more powerful telescopes, we know these objects to be huge far-away galaxies of stars, like the galaxy in which our Sun is a very small member.

When it was first proposed in 1796, this Nebular Hypothesis of Laplace seemed to account for most of the facts that had been observed about the solar system. However, the more astronomers learned about the movements of the members of the Solar System and the behavior of great masses of gas, the less sure they were that this hypothesis would explain all the facts. They began to doubt that the mechanics of the system would work. They said that even if rings would form, they could not condense to form planets, with or without moons. The spinning of the present system, they argued, could not have developed within the nebula. It must have been caused by some force outside.

About 1905, two scientists at the University of Chicago, Thomas C. Chamberlin, a geologist, and F. R. Moulton, a mathematician, began work on what they thought was a better theory. They developed what they called the Planetesimal Hypothesis. The idea was some-

thing like this: long ago our Sun was a star without any planets. As it traveled through space it chanced to come close to another star. The attraction of the other star, as it went by, pulled great waves of gas away from our Sun and set them spinning like a pinwheel. Perhaps the other star also had part of its gas set spinning about it, and now may have a family of planets and moons, somewhere in space. At any rate, the knots of hot gas that were set spinning around what remained of our Sun condensed into solid particles of various size. These they called planetesimals. Through their mutual attraction they gathered into larger knots, as they spun along in their orbits around the Sun. The larger masses of gas and particles in time swept up or drew to themselves all the planetesimals in their paths, grew in size, and became the planets. Some masses, which these could not absorb, were captured and became the moons, circling in orbits around the planets. In this model, Saturn's rings are still in the planetesimal stage, small particles forming distinct bands one beyond the other, extending out to a total diameter of 171,000 miles, but only about 1 mile thick. Between Jupiter and Mars the planetoids (asteroids) did not consolidate into very large masses, but remained small, up to several hundred miles in diameter, each following its own orbit around the Sun.

This planetesimal theory has been modified over the subsequent years. Sir James Jeans, in his "tidal theory," suggested that, instead of the material which was pulled away from the sun forming a myriad of  planetesimals spiraling out from both sides of the sun, it formed one huge, spindle-shaped mass of glowing gas extending out from one side of the Sun. Jupiter and Saturn, the largest planets, were formed near the wide center of the spindle. The inner planets were formed in the slender, inner end of the spindle, where the material, not having been thrown so far, was heavier and denser. The two outer ice giant planets again are smaller and were formed in the narrowing outer end of the spindle of gas. Harold Jeffreys then suggested a further modification that the star and our ancestral sun actually hit, or at least sideswiped, each other, rather than causing tidal disruptions by their close approach. In this scheme, only a small proportion, (about 1/700th) of the Sun's mass had to be pulled away to form the planets, so no very great disruption need have taken place.

However, today astronomers realize that the probability of such an encounter between two stars like the

Sun in the local solar neighborhood is negligible. More-over, we know from recent imagery of star-formation regions in our Milky Way, particularly with infrared (heat-sensitive) cameras, that dusty disks and rings surround many stars (for example, Vega and Fomalhaut) in regions where stellar collisions are improbable. There are over 500 such objects in the Great Nebula of Orion alone! Thus, it would seem that the formation of such protoplanetary disks is a natural consequence of star formation, and it is unnecessary to account for the origin of the Solar System as a low-probability process. How such a disk might become a family of planets is now explained by a combination of both the Nebular Hypothesis and the Planetesimal Hypothesis without resorting to a chance stellar encounter.

What begins as a spherical cloud ends up as a flat disk of gas and dust, spinning in a particular direction in space. This is because gravitational collapse proceeds more quickly along the rotation axis of the cloud, where there is little inertial motion ("centrifugal" force) to overcome. Thus the theory accounts for the uniform direction of revolution and the alignment of planetary orbits. The central regions would be spinning fastest of all just like an ice skater spins faster when she brings her arms toward her body. (This is the law of conservation of angular momentum). In the hotter, inner regions of the solar nebula, the volatile and icy material would be vaporized; only dense, iron-rich compounds could remain solid. Rocks and dust would collide and cling together by electrostatic forces. Later, clumps of matter would gravitate to form planetesimals of 50 miles in diameter, which would then begin to sweep up remaining debris by gravitational accretion. Larger bodies would begin to grow, sweeping up matter in each orbit. This process can lead to the formation of 4-6 rocky iron-rich planets in about 100 million years, according to computer simulations. The formation of the larger Jupiter-sized planets would also begin with accretion of planetesimals. Beyond the asteroid belt, temperatures in the young solar nebula would be cold enough that ice-coated dust and volatile elements were not vaporized. The outer planets would accrete large envelopes of frozen hydrogen-rich matter on top of their rocky cores. After 10-15 Earth masses were accumulated, the outer protoplanets would begin to pull

in surrounding gases moving at low velocities, and would acquire massive atmospheres. When the central nebular mass became hot enough in its core for nuclear fusion to "turn on," a gigantic solar wind of atoms would sweep the young Solar System clean of all the excess gas, leaving behind rocky debris to impact all the planets for a half-billion years or so until it was all swept up. The record of this early bombardment is the high density of craters on the surfaces of Mercury, Mars, and the Earth's Moon, not to mention the icy surfaces of the moons of the four giant outer planets. On Earth and Venus, many craters have been erased by volcanic and erosional activity. The atmospheres of the outer planets apparently absorbed most of these early impactors, but the fact that the rotation axis of Uranus is tilted on its side suggests that even the giant planets did not escape the disruptive effects of the early bombardment stage.

This modern Protoplanet Hypothesis (developed by V. Safronov, R. Greenberg, G. Wetherill and others) for the origin of the Solar System accounts for many observations about the Sun and its family of planets which would otherwise be difficult to explain. It accounts for why (1) all planetary orbits lie in nearly the same plane as the Sun's rotational equator; (2) why the planetary orbits are almost circular; (3) why the planets revolve (and almost all rotate) in the same direction, (4) why planetary moon systems look similar to the Solar System as a whole, (5) why cometary orbits outline a large spherical cloud, and (6) why the inner planets differ in composition from the outer planets, with the density generally decreasing outward. Bodies that formed at higher temperatures in the young solar nebula (closer to the Sun) have incorporated into them a lower proportion of compounds that melt or vaporize at low temperatures, such as the inner planets. This also explains why meteorites differ in composition from lunar and Earth rocks.

It is now understood that the asteroids, having densities intermediate between the inner and outer planets and similar to those of meteorites, are likely fragments of planetesimals which never accumulated into a larger body. This is because Jupiter's gravity kept them "stirred up," moving at higher velocities with respect to each other. Such high relative velocities would have resulted in these objects smashing into each other. Only the fragments moving at low relative velocities in nearly circular orbits remain today. Many of the fragments undoubtedly found their way into the inner Solar System. Some of these bodies survived the descent through Earth's atmosphere to become meteorites. Just as there is a ring of leftover rocky debris orbiting between Mars and Jupiter, there is also a ring of icy debris orbiting outside the orbit of Pluto, called the Kuiper Belt. Jupiter and Saturn have also deflected some icy planetoids into the inner Solar System, resulting in the periodic appearance of comets. On every passage around the Sun, a comet loses 0.1-1% of its mass, so after about 100-1000 trips, there is nothing left but dust and debris. This material continues to circle the Sun along the comet's orbit, and if the Earth passes through the path of a former comet, the result will be a meteor shower. Thus, meteor showers are associated with icy comets, while meteorites are generally associated with rocky asteroids. However, an individual meteoroid could be either rocky or icy, and may originate within either a larger cometary or asteroidal body.

One observation which is inconsistent with the modern Protoplanet Hypothesis is the Sun's very slow spin rate (25-36 days) compared to that of the planets (10-24 hours, except for Mercury and Venus, which have been slowed down by tidal forces). This can be explained if excess charged (ionized) gas ejected by the magnetic Sun in its pre-fusion stage slowed its rotation, such as a tennis ball losing spin faster when wet. Recent observations of other young protostars supports this contention.

Many details concerning the planets, their moons, comets, asteroids, meteors and meteorites can be found in numerous good books on astronomy. The planets are not shown on the star charts; they move about rapidly. However, they are always located in the zodiac constellations, not very far from the ecliptic, as shown on the Mercator chart (pg. 32-33), where there are only a few bright stars with which they might be confused. If there is a bright object in this zone of the sky, not marked on the star chart, it probably is a planet.

The Planets and their Moons

The age of solar system exploration began in the 1960s and has continued up to the present. As a result, the subject of planetology has been revolutionized. Missions to every planet have generated a wealth of information. Close-up views of all the major planetary satellites have been obtained, revealing that some of these are spectacular planetary worlds in their own right. A brief voyage of rediscovery follows.

Earth, of course, is the only planet in the solar system known to harbor life, a fact that is likely connected with the presence of liquid water on its surface. The crust of the Earth is broken up into large plates, which drift on top of a partially molten layer 240 miles below. The plates butt up against each other, generating frictional heat that produces chains of volcanoes and earthquakes at plate boundaries. As the Pacific plate drifted northwest over a hot spot in the Earth's upper mantle, a chain of volcanic islands known as Hawai'i were formed. The Earth's crust is comparatively cool due to the presence of water, and thus ancient craters are repaved and erased by means of plate movement and recycling of volcanic rock. This dynamic process of Earth changing its surface quickly seems to be unique among the other rocky planets or moons orbiting other planets.

Water was once thought to be highly unique to Earth in the Solar System. However, astronomers now view the Solar System differently. The Moon, for one, was once thought to be dry and unchanging. Not visited by humans since the Apollo astronauts did between 1969 and 1972, recent robotic probes have found water the earlier astronauts could not, even by carefully studying the 800 pounds of basalt and anorthosite they brought back to Earth. The Moon has tremendous reserves of water hidden beneath its seemingly dry surface. Some of the Moon's water hides as ice along deeply shadowed crater walls, perpetually hidden from direct sunlight. This water is able to slowly migrate around to different locations on the Moon in lunar water cycles, somewhat like Earth's water experiences. Perhaps even more surprising, Mercury, the planet orbiting closest to the Sun, also hides water ice from the Sun's energy along its the walls and steep cliffs.

The tiny planet Mercury has notoriously been difficult to study. For one, it is always close to the brilliantly shining Sun, making telescope observations difficult. For another, this planet has a curious dynamic configuration: it rotates (spins) on its axis 3 times for every 2 trips around the Sun, so that one Mercurian day equals 176 Earth days. Mercury's average density is 5.4 times the density of water, so it must have a large iron core extending out about 75% of the way to the surface; this is a larger proportion than for any other planet. Moreover, Mercury exhibits an unexpected magnetic field, but one that doesn't seem to emanate from its overly large iron core. As a planet, Mercury, still holds many mysteries, despite a detailed analysis from NASA's Mercury Messenger space probe that orbited it from 2011 until it crashed onto the surface in 2015.

Venus is supposedly the Earth's twin, but as demonstrated by Pioneer Venus Orbiter in 1978 and by the Magellan Orbiter (1990-1994), the planet is much more an opposite, than a twin. The surface temperature is 900°F, even hotter than Mercury's sun-facing side, due to a thick atmosphereic blanket of heat-holding carbon dioxide, which is at a pressure of 90 atmospheres near the surface. The atmosphere is devoid of water, and corrosive sulfuric acid clouds permanently shroud the planet's surface from view in visible light. Radar imaging by Magellan—which is able to penetrate the thick atmosphere of Venus to give a view of its hellish surface—reveals a world covered with solidified lava flows, probably the result of catastrophic volcanism triggered by climate change 800 million years ago. At that time, the entire surface was repaved by magma; huge volcanic structures are seen dotting the landscape. Channels of runny lava, now hardened, flowed for thousands of miles across the barren surface. There seems to be little activity at present, however. Why the volcanoes stopped erupting millions of years ago is still a mystery.

Mars, in stark contrast to Venus, has features resembling dried river beds or channels in abundance, first detected by the Mariner 9 spacecraft in 1971. Near the equator of Mars is a plateau five miles high on which a quartet of enormous volcanic structures stand; the largest of these is Olympus Mons, 15 miles high and 700 miles across. The island of O'ahu would easily fit into its caldera. Eruption of magma probably ceased 150 million years ago. Adjacent to the plateau is a giant chasm 3000 miles long, as wide as the United States. Despite Mars' popularity as the home of science-fiction alien invaders, there is no evidence for any forms of life in the Martian soil, apparently sterilized of organic material by harmful ultraviolet light. Today, Mars has an air pressure level so low that liquid water can no longer stably exist at the surface, so Mars must have had a denser atmospheric blanket in the past capable of keeping water flowing—perhaps 7.5 atmospheres of carbon dioxide gas.

No planet has given exploring planetary space probes more challenges than Mars. Starting in 1960, more than 50 space probes have been launched toward Mars, but only about 1/3 have been successful. The surface, as revealed by the Viking 1 and 2 missions in 1976, and by Pathfinder in 1997, is dry and dusty red with a pink sky. Global dust storms lasting for weeks are common, and large ice caps of frozen carbon dioxide surrounded by dune fields are visible from orbiting spacecraft such as the Mars Global Surveyor (1997-2006) and Mars Odyssey (2001-2007).

Jupiter and its satellite system have been explored by Pioneer 10 and 11 in 1973, Voyager 1 and 2 in 1979, by the Galileo spacecraft (1995-2003), and by the Juno mission that arrived in 2016. Jupiter is the largest planet in our Solar System, almost 11 times wider than the Earth and 318 times more massive. Rather than being comprised mostly of silicon-rich or iron-rich rock, the outer planets are made up mostly of hydrogen and helium.

Although these are in gaseous form in the atmospheres of the giant Jovian planets, gas becomes liquid under millions of atmospheres of pressure deep inside. Jupiter's turbulent clouds are extremely colorful; the most obvious feature is the Great Red Spot, a storm system twice as large as the Earth itself. Jupiter whirls around on its spin axis once every 10 hours, spreading the clouds into bands, and these can even be seen with a small telescope.

The planet has a multitude of small moons or satellites, but the four discovered by Galileo in 1609 are unique and dynamic worlds unto themselves. Io and Europa are about the size of Earth's moon, while Ganymede and Callisto are about the size of Mercury. The latter two are icy worlds with many shallow impact craters. Europa is also ice-covered; its surface resembles polar terrain on Earth, but the ice is crisscrossed with dark cracks, and there are few apparent craters. By carefully monitoring how it spins, scientists are convinced that Europa hosts an ocean of liquid water underneath its icy exterior. Moreover, Europa is actively being repaved not with magma, but with moving ice sheets.

Orbiting closer to Jupiter than the other moons, Io is the most volcanically active world in the Solar System, with plumes of sulfur ejected 75 miles high. Io's quickly changing surface resembles a cheese pizza, with a range of striking colors due to sulfur deposits of various temperatures. This world is being resurfaced on a short timescale (months) due to molten rock flowing across its surface, melted by internal heat generated as Io flexes. Io is caught in a tug-of-war between Jupiter and the other three Galilean satellites, and it reacts to the changing gravitational attraction by heating its internal rocks.

The Saturnian system, consisting of rings and more than 60 moons, has been explored by Pioneer 10 and 11 (1974), Voyager 1 and 2 (1980-1981), and has been under constant surveillance of the Cassini–Huygens spacecraft since 2004. Saturn is the least dense planet in our solar system. If there was an ocean of water large enough, Saturn would float! It is less colorful than Jupiter, probably because its wind speeds (up to 1000 miles per hour) are higher. The planet's most notable feature is of course the ring system, first revealed in glorious detail by the Voyager flyby missions.

The rings are composed of millions of chunks of water-ice ranging from dust-sized to mountain-sized. The  ice chunks have been shaped by gravity into a distribution that resembles a phonograph record, but it has been clear for over 100 years that the rings are not a solid mass. The ring debris is likely the remains of one or more icy satellites that came too close to Saturn and were smashed by tidal forces. Saturn has a large number of icy satellites, most of which show evidence of an ancient phase of impact bombardment by comets. Two of the satellites are especially interesting. Iapetus has one entire hemisphere covered with a coal-black deposit whose origin is uncertain. Titan, second only to Ganymede in size among satellites (both larger than Mercury), is shrouded by a thick atmosphere of nitrogen and methane gas. This remarkable satellite may have lakes of liquid petroleum on its surface.

Uranus and Neptune have only been visited by Voyager 2 in 1986 and 1989, respectively. Uranus is unique in that it is tilted on its side, probably the result of a collision with a large planetoid early in its history. Its ring system must have formed after that event because it is nearly vertical to the plane of the planet's orbit. The atmospheres of Uranus and Neptune are strikingly blue because of a 2% abundance of methane gas. Probably much of the interiors of these two blue balls is composed of liquid water at high pressure, as might be found at the bottom of Earth's deepest oceans. Neptune had a storm system (called the Great Dark Spot) in 1989 when Voyager flew by, but since then the storm system has disappeared! Both worlds have families of icy satellites and narrow rings, but two members of these families stand out. Miranda, an icy sphere 300 miles across which orbits Uranus, shows a tortured landscape of icy ridges and fractures uncharacteristic of such a small object. Cliffs five miles high are seen in one area. Was Miranda subjected to extreme tidal effects or was it blown apart by a collision and then reassembled into a pile of rubble? Triton, which orbits Neptune, has a polar cap of pink methane ice which occupies half the satellite. On this polar cap are dark streaks identified as active geysers, spewing liquid air (nitrogen) five miles above Triton's surface. Theorists were hard pressed to come up with an explanation for a source of heat beneath the crust of a world only 40 degrees above absolute zero!

Finally, in the frigid outer reaches of the Solar System, we find tiny Pluto, classified as a dwarf planet. Discovered in 1930, it has not completed even one orbit since its discovery. Pluto has an interesting satellite, an icy rubble pile half the planet's diameter known as Charon. Curiously, Charon orbits Pluto in exactly the same time as a day on Pluto (6.39 Earth days). Thus, an observer on Pluto's surface would see Charon suspended in the sky, always at the same altitude above the horizon. Pluto is the largest of a family of planetoids known as Plutinos. They are probably survivors from an early period of bombardment in the history of the Solar System. Triton may be another survivor from this period that was subsequently captured by Neptune's gravity. Pluto is certainly not an escaped satellite of Neptune, as had been thought previously, since it never comes close to the blue giant even though their orbits actually cross.

STATISTICS ABOUT THE SOLAR SYSTEM

Name	Mean Distance From Sun Miles	Mean Diameter Miles	Mass* Earth = 1	Volume Earth = 1	Stellar** Magnitude (see note)	Period of Revolution (Sidereal year)	Period of Rotation (Sidereal day)	No. of Moons
Mercury	35,980,000	3,032	0.0553	0.0560	-1.9	87.97 days	58.65 days	0
Venus	67,230,000	7,521	0.815	0.854	-4.4	224.7 days	-243.0 days (R)	0
Earth	92,960,000	7,926	1.00	1.00	------	365.26 days	23h 56m	1
Mars	141,600,000	4,222	0.108	0.151	-2.0	687.0 days	24h 37m	2
Asteroids***								
Jupiter	483,600,000	88,850	318	1,410	-2.7	11 yr. 315 days	9h 55m	67
Saturn	888,200,000	74,900	95.2	844	0.67	29 yr. 167 days	10h 14m	62
Uranus	1,784,000,000	31,760	14.5	64.4	5.5	84 yr. 4 days	-17h 54m (R)	27
Neptune	2,799,000,000	30,750	17.1	58.4	7.8	164 yr. 288 days	16h 7m	14
Pluto	3,674,000,000	1,466	0.0022	0.0066	15	248 yr. 197 days	-6.387 days (R)	5
Sun	------	864,000	332,830	1,300,000	-26.7	------	25-36 days	------

* The Earth's mass is 6,600,000,000,000,000,000,000 tons (5,976,000,000,000,000,000,000,000 kilograms). Its density is 5.515 times that of water. To calculate the mean density of each planet in terms of the density of water, divide the mass of the planet (in Earth masses) by the volume of the planet (in Earth volumes), and then multiply that result by the Earth's mean density of 5.515 (in terms of water's density).

** The stellar magnitude of the "outer planets" is that of opposition; of Mercury and Venus at elongation.

*** The asteroids move in individual orbits, from 135,000,000 to 488,000,000 miles from the Sun, in periods from 1¾ to 12 years. The diameters of the 20,000 numbered asteroids that have been measured are from 0.5 to 583 miles. Only 16 asteroids have a diameter greater than 150 miles.

(R) means retrograde rotation, which is from east to west; the other planets (including Earth) rotate prograde, or from west to east.

Chapter 5
Stars, Galaxies, and Cosmology

How many stars are there?

The number of stars in the entire universe must be very great, but only about 7,000 can be seen by the unaided human eye. On a clear, moonless night, a person with good eyesight could count about 2,500 stars, visible at any one time. The more astronomers magnify the heavens the more stars are seen on photographs or images. The following table gives the approximate number of stars in each magnitude and by what means they can be seen. It is difficult to estimate the number of stars fainter than 15th magnitude.

Magnitude	Number of stars	How visible
1st	20	
2nd	58	
3rd	82	
4th	530	
5th	1,600	
6th	4,800	
Total	7,090	visible to the unaided human eye
7th	15,000	through an opera glass
8th	46,000	through binoculars
9th	140,000	through a small telescope
10th	380,000	a larger telescope
11th	1,000,000	about a 4 inch telescope
12th	2,600,000	about a 6 inch telescope
13th	5,000,000	about an 8 inch telescope
14th	13,000,000	about a 10 inch telescope
15th	27,000,000	about a 12 inch telescope
16th	157,000,000	a very large telescope

The Sun and all of the stars that we can see with our naked eyes, most of those which we can see through small telescopes, and many which we cannot see, belong to one huge discus-shaped group, which we call our galaxy. We know that our galaxy has the general shape of a pocket watch or of a disk, because it has been imaged directly by satellite detectors such as the Cosmic Background Explorer. Our Sun and its system of planets, in-cluding the Earth, are located out toward one of the edges of this flat disk. When we look the long way of the galaxy we see that the great number of stars in that direction seem to form a band of hazy light, which we call the Milky Way. There are at least 100 billion stars in our galaxy. We could not see all of them, no matter how much we magnified them, because many are hidden behind curtain-like masses of gas and small "dust" particles of solid material. Such a mass is called a nebula. Some nebulae reflect the light from nearby stars and have a hazy brightness; others are dark, but we can see them because they obscure portions of the Milky Way that lie beyond, appearing as dark blotches against the bright hazy background. Our galaxy is estimated to measure about 100,000 light years across. More than a million tiny patches of light, upon being imaged through the largest telescopes, have turned out to be great swarms of billions of stars as well. These are galaxies external to our Milky Way system.

The distances to the stars

If two persons on the earth, say two or three thousand miles apart, were to look at the moon at the same instant, they would see it in somewhat different directions. Using the distance between the observers (A and B in Figure below) as a base and the two angles resulting from difference of direction, it is possible by trigonometry to calculate the distance to the moon. This same method can be used to measure the distance to the Sun, and the nearer planets. The stars are so far away that there would be little difference of direction from the ends of such a short base line. However, astronomers can use a longer base. They can observe the direction of a star, or more practically photograph its position with reference to more distant stars from opposite sides of the Earth's orbit around the Sun. A photograph taken in January might show that the nearer star was on one side of the

Measuring the distances to the stars

more distant star; one taken in July might show that it lay on the other side of the distant star (x and y on the diagram). This is the method of stellar parallax. The closer the star is to Earth, the larger will be the apparent shift in position. If half the total shift, referred to as the parallax angle, equals 1 arcsecond (1/3600 of a degree), then the star is 206,265 times the Earth-Sun distance away. We define this distance to be 1 parsec. If the star showed a parallax angle of 0.5 arcseconds, then the star would be 2 parsecs away; if the parallax angle were 0.05 arcseconds, then the star would be 20 parsecs away, and so on.

Much greater distances have been determined by another method. Certain stars, called Cepheid variable stars, show sudden increases of brightness at regular intervals, as if they underwent great pulsations. These stars are expanding and contracting, much like breathing in and out. It has been found that there is a close relationship between the period of time separating maxima in brightness and the actual sizes of these stars, and this gives a measure of their absolute magnitudes (the star is more luminous when it is larger). The longer the period, the brighter and more luminous the star. If a Cepheid variable in a faraway cluster of stars is observed to have a certain interval between pulsations, it is possible to determine its absolute magnitude. Comparing its apparent magnitude with that of another Cepheid variable, with known absolute magnitude and known distance (due to parallax, for example), astronomers can calculate the distance to the new Cepheid variable, and hence of the cluster or galaxy in which it is located. As long as the light changes can be seen, identifying the star as a Cepheid variable, distances to the system in which the star resides can be determined, even up to millions of light years.

Star magnitudes

The ancient Greek scholars divided the stars into groups called "magnitudes." They called the brightest stars "first magnitude" and the faintest (which they could see, having no telescopes) "sixth magnitude." When astronomy became a more exact science, astronomers gave a more precise meaning to the term magnitude. Astronomers found that first magnitude stars were about 100 times brighter than those that could just be seen with the naked eye (6th magnitude). That made one magnitude about two and a half (2.512) times brighter than the next. First magnitude stars are 2.512 times brighter than second; second are 2.512 times brighter than third, and so on. A few of the brightest stars were found to be so much brighter than others, which had also been called "first magnitude" that these had to be called "zero magnitude." Two stars, brighter than zero magnitude, were given "negative" values; Canopus is magnitude -0.62, and Sirius, is -1.44, almost one and a half magnitudes brighter

than zero magnitude and two and a half magnitudes brighter than first. Only astronomers speak about hundredths of a magnitude, although in comparing one star with another through a telescope, as in studying variable stars, the human eye can estimate differences of a tenth of a magnitude without difficulty.

Apparent visual magnitude is defined as the visual magnitude as seen from Earth. Absolute visual magnitude is defined as the apparent visual magnitude as seen from a distance of 32.6 light years (equal to 10 parsecs), and is a better measure of a star's true luminosity (light energy radiated in each second). The difference between these two quantities (apparent magnitude minus absolute magnitude) becomes a larger positive number if the star is farther away from Earth. If this difference is 0 magnitudes, the star is 10 parsecs away; if this difference is 5 magnitudes, the star is 100 parsecs away; if this difference is 10 magnitudes, the star is 1000 parsecs away, and so on. This follows since the intensity of light measured drops off rapidly as the light source distance increases. Stars with negative absolute magnitudes would be as bright, or brighter than the planets, if they were as close as 10 parsecs, or 32.6 light years. These are stars that are intrinsically luminous, i.e., they radiate a great deal of light energy. Luminous stars tend to be seen at great distances, and so many of the twenty brightest stars are not very close. The brightest star (with the most negative apparent magnitude) Sirius is bright because it is close. Yet the second brightest star, Canopus, is a long way off in comparison. Canopus is 36 times as far away as Sirius. Sirius is 26 times as bright as our Sun would look if they were both 32.6 light years away, but Canopus is 14,000 times as bright as our Sun if they were both 32.6 light years away. The table on page 46 gives this kind of information about the twenty brightest stars in the sky, those of the first magnitude and brighter.

Variable stars

Many stars vary in brightness. Even the light from our Sun is not constant, partly because of the increase and decrease in number of sunspots. Astronomers recognize at least five different kinds of variable stars: (1) Binaries are pairs of stars that revolve about their common center of gravity. If both are bright, when side-by-side, the maximum amount of light is given off. If one passes in front of the other, part of the light may be eclipsed. Algol, in Perseus, is a famous example of an eclipsing binary. (2) Cepheid variables (so named because several were found in the constellation of Cepheus), as noted previously, change their brightness rhythmically, as if pulsating. Mira, a red giant star in Cetus, varies from second magnitude down to almost out of eyesight, and is the best-known example of a long-period (Mira) variable. (3) Certain irregular variables may pass behind

STATISTICS ABOUT THE TWENTY BRIGHTEST STARS

Rank	Arabic Name	Hawaiian Name	Apparent V Magnitude	Distance Parsecs	Distance Light years	Absolute V Magnitude	Spectral Classification	Radial Velocity in km/sec	Proper Motion in arcsec/yr	Luminosity (Sun = 1)	Multiplicity or Variability
1	Sirius	A'a	-1.44	2.64	8.61	1.45	A1 V	-9	1.34	26.1	(binary)
2	Canopus	Keali'iokona	-0.62	96.0	313	-5.53	F0 I	+21	0.031	14,000	
3	Alpha Centauri	Kamailehope	-0.28	1.35	4.4	4.34	G2 V	-25	3.71	1.77	(triple)
4	Arcturus	Hokule'a	-0.05	11.3	36.7	-0.31	K2 III	-5	2.28	190	
5	Vega	Keoe	0.03	7.76	25.3	0.58	A0 V	-14	0.035	61.9	
6	Capella	Hokulei	0.08	12.9	42.2	-0.48	G8 III	+30	0.434	180	
7	Rigel	Puanakau	0.18	260	860	-6.69	B8 Ia	+21	0.002	700,000	(binary)
8	Procyon	Puana	0.40	3.49	11.4	2.68	F5 IV-V	-4	1.26	7.73	(binary)
9	Betelgeuse	Kauluakoko	0.45	197	643	-5.14	M2 Iab	+21	0.029	41,000	(variable)
10	Achernar	Kalanikauleleaiwi	0.45	44.2	144	-2.77	B3 IV	+19	0.097	5250	
11	Beta Centauri	Kamailemua	0.61	107	350	-5.42	B1 II	-12	0.042	86,000	
12	Altair	Humu	0.76	5.15	16.8	2.20	A7 IV-V	-26	0.661	11.8	
13	Acrux	Kamolehonua	0.77	99	320	-4.10	B0 IV	-11	0.035	25,000	(binary)
14	Aldebaran	Kapu'ahi	0.87	20.0	65.1	-0.63	K5 III	+54	0.199	370	(binary)
15	Spica	Hikianalia	0.98	80.4	262	-3.55	B1 V	+1	0.053	25,000	
16	Antares	Lehuakona	1.06	185	604	-5.28	M1 Ib	-3	0.025	37,000	(binary)
17	Pollux	Nanahope	1.16	10.3	33.7	1.09	K0 III	+3	0.627	46.6	
18	Fomalhaut	Kukaniloko	1.17	7.70	25.1	1.74	A3 V	+7	0.368	18.9	
19	Deneb	Hawa'iki	1.25	802	2600	-8.73	A2 Ia	-5	0.002	320,000	
20	Beta Crucis	Kauli'a'ama	1.25	108	353	-3.92	B0.5 III	+20	0.050	34,000	

Notes: Data from this table was compiled from the Hipparcos General Catalogue.

Radial velocity with (-) sign is toward the Sun and Solar System. Proper motion is tangential motion across sky in one year (3600 arcsec in 1 degree).

Absolute magnitude is the apparent magnitude a star would have at a distance of 10 parsecs or 32.6 light years.

Luminosity classes: Ia, Iab, Ib (supergiant), II (bright giant), III (giant), IV (subgiant), V (main sequence dwarf), VI (subdwarf), VII (white dwarf)

Spectral classes and corresponding absolute surface temperatures: B0 (30,000 Kelvin), A0 (10,000 K), F0 (7,500 K), G0 (6,000 K), K0 (5,000 K), M0 (4,000 K), M9 (3,000 K); each spectral type is subdivided into 10 classes from 0-9 with intermediate temperatures. Hottest stars are blue; the coolest are red.

masses of dark gas or in some other way have their light reduced at irregular intervals. (4) Small red dwarf stars are often variables because their photospheres are dominated by large "starspots", as our Sun is in a mild way. (5) A "nova" may flare up very brightly and then slowly fade away as the result, it is thought, of some tremendous explosion or nuclear-fusion reaction on its surface. This is usually attributed to the transfer of mass from a companion in a binary system.

Weighing the stars

By studying the motion in their orbits of members of a system, such as the moons around a planet, or the planets around the Sun, it is possible to calculate their relative weights. Actual weights can be measured by relating them all to the Earth, the mass and size of which are known. Applying these same rules of the pull of gravity to systems of stars, it is possible to weigh them also. Whenever astronomers carefully observe a pair of binary stars they can weigh both of them. For example, the closest star to the solar system, Alpha Centauri, is a triple system. One member of the trio has been found to weigh 1.14 times as much as our Sun, and another member 0.97 times as much. Although both weigh about the same, there is a great difference in the amount of light they give. The larger gives 1.12 times as much light as our Sun, showing that it is very much like our Sun, but the smaller gives only 0.32 times as much light as the Sun. There is an even greater difference between Sirius A and its binary companion Sirius B. Sirius A weighs 2.45 times as much as our Sun and gives 26.1 times as much light; its companion weighs 0.85 as much as the Sun, but gives only 0.0026 as much light. Thus Sirius B is a very dim star; it is undoubtedly small, but very dense. Knowing the weight and diameter of a star, it is possible to calculate the density. The density of Sirius B must be millions of times denser than water, and yet its size is approximately that of the Earth. We call such a small hot star a white dwarf. It is likely the burned-out core of a star once like our Sun is now.

The color, temperature and movement of stars

We notice that stars have different colors: red, orange, yellow, white, and blue. The color is related to the temperature at the surface. Astronomers are not only able to measure the temperature, they are also able to tell the chemical composition of the gases by using a spectroscope (or spectrograph) to study the light that comes from the star. All of the elements found on the Earth have been found on the Sun.

Small shifts in the position of the lines of the spectrum, seen through a spectroscope, also tell whether the star is coming toward us or moving away from us. The amount of such shift toward the red end of the spectrum gives a measure of the speed with which the star is moving away from us; the amount of shift toward the violet end is in proportion to the speed of approach. This is called the Doppler effect. Such measurements give the radial movement of stars away from or towards the Earth and the rest of the Solar System. A study of the radial velocities of many stars shows that the Sun and its entire system is moving through space in the general direction of the constellation of Lyra or Hercules at about 12 miles a second. Comparisons of photographs made of stars from time to time through large telescopes show minute shifts of position of some nearby stars as compared with fainter, and more distant stars. This is called proper motion. When the distance to the nearby star is determined from parallax measurements, the speed of "transverse movement" across the line of sight, called tangential velocity, can be determined. Figures for these radial and tangential velocities, in kilometers per second, are given in the tabulation of the twenty brightest stars.

The life cycle of a star

After studying the size, temperature, mass, luminosity and spectral features (measure of composition) of many stars, astronomers find that they fall into groups, or types. There are red supergiant stars (such as Betelgeuse) 1000 times larger than our sun; and small red dwarf stars only 0.1 the size of our sun. There are also dense white dwarf stars (such as Sirius B) which are the size of the Earth. The range in stellar masses is from about 0.08 solar masses (faint red stars of temperature 3000 degrees absolute) to about 60 solar masses (bright blue stars of temperature 30,000 degrees absolute). Stellar luminosities range from 0.0001-1,000,000 in solar units (1 solar unit is about 400 trillion trillion Watts) so the range in luminosity is much greater than the range in mass. Astronomers find from binary star data that there is a clear relationship between the masses and luminosities of most stars. The mass-luminosity relation states that the more massive the star, the more luminous it is. This is such a fundamental relation that the mass is the most basic property of a star, and determines its luminosity, temperature, size, and also evolutionary history.

The luminous energy of stars like the Sun is generated by nuclear fusion at the Sun's core, where the temperature is 15 million degrees absolute. At these high temperatures, four hydrogen nuclei (protons) fuse into a single helium nucleus. The helium nucleus is 0.7% less massive than the total mass of the four protons, so the difference is transformed into light energy. By this process, the Sun is converting 660 million tons of hydrogen gas into helium gas every second; it is actually radiating away 4 million tons each second, yet because the Sun has such an enormous mass (333,000 times that of Earth), fusion of hydrogen into helium (called hydrogen "burning") can continue for about 10 billion years for a star of the Sun's mass. Curiously, more massive stars have shorter hydrogen-fusion lifetimes; their core temperatures are much higher, so the nuclear fusion uses up the hydrogen much faster. They are like brighter "candles" that "burn out" faster. A star 10 times more massive than the Sun uses up its hydrogen fuel 100 times faster, which corresponds to a hydrogen-fusion lifetime of 100 million years; a star 50 times more massive than the Sun uses up its hydrogen fuel in only 4 million years. There are small changes in the Sun's heat and light output, but there has not been any overall marked decrease during geologic time, so we can expect the Sun to keep on giving heat and light to its planets for many billions of years to come. However, when all the hydrogen gas in the Sun's core has been converted into helium gas, a drastic change in its structure will take place. In about 5 billion years, the Sun will bloat up, becoming a red giant (like Aldebaran or Arcturus), and will engulf Mercury and Venus; here on Earth, the oceans will vaporize and all life will likely be extinguished.

As hydrogen is transformed into helium in a star's core, the core shrinks gradually, drawing the matter to ever-increasing temperatures. At the end of the hydrogen-fusion phase, the core shrinks on a shorter time scale (thousands of years compared to billions of years) because there is no energy source within it to stave off the relentless force of gravity. As the temperature in the core reaches 100 million degrees, a new reaction begins, the fusion of 3 helium nuclei into a nucleus of carbon ("helium burning"). As helium is transformed into carbon (and some oxygen) in a star's core, the star swells up even more and becomes an even bigger red giant. If the star is a low-mass star like the Sun, pulsations in the giant star's outer layers are triggered at this stage, and the star becomes a Mira variable, expanding and contracting about once each year. During each pulsation, part of the star's outer envelope is ejected. Shells of dust form around the star which are pushed away on each cycle until the low-density hydrogen envelope is almost totally removed. Our low-mass star quickly evolves (in about 1000 years) into a hot blue object (the core of the star) surrounded by glowing shells of gas that we call a "planetary nebula" (this inappropriate name was coined by William Herschel, since their turquoise hue reminded him of Uranus). Eventually the shells of gas dissipate outward (in less than 50,000 years). The remnant core has insufficient mass to force contraction to more advanced stages of nuclear fusion. All it can do at this stage is cool down until it becomes an object we call a white dwarf. The Indian astrophysicist S. Chandrasekhar showed in the 1920's that such a star must have a mass less than 1.4 times the mass of the Sun after the planetary nebula stage.

High-mass stars (more than about 8 times the mass of the sun) end their lives differently. This is because higher central temperatures are reached in their cores. After the helium-fusion stage ends, core contraction increases the temperature to more than 600 billion degrees, at which point carbon fusion into heavier elements commences. At each stage, the "ash" of nuclear fusion becomes the "fuel" of the next, as the core temperature rises into the billions of degrees. Carbon is turned into oxygen, neon, or magnesium; oxygen is fused into silicon, phosphorus, or sulfur. The ultimate end product of this sequence is iron, but that is where nuclear fusion comes to a halt. As lighter elements are fused into heavier elements, the stability of the product nucleus increases; that is, until a core of iron is formed. Iron is so stable that no more energy can be released from fusion. When 1.4 solar masses of iron have accumulated in a massive star's core, the core can no longer support itself against the relentless force of gravity, and rapid core compression occurs. Iron nuclei are broken down into protons and electrons that squeeze together to form a 6-mile-wide sphere of neutrons. The process releases copious numbers of nearly massless particles called neutrinos, which deliver enough energy and momentum to the surrounding gas shells to disrupt the entire star in a massive explosion called a supernova. The energy released in a supernova is equivalent to the visible energy of the Sun radiated over its entire 10-billion year hydrogen-fusion lifetime! Although the above scenario for the death of a massive star is theoretical, the theory is supported by systematic observations of neutrinos from Supernova Shelton (SN 1987A). This supernova event observed in 1987 was the explosion of a massive blue supergiant in a small galaxy (called the Large Magellanic Cloud) that orbits our Milky Way.

The remnant neutron star that survives the supernova explosion must be spinning extremely rapidly as a result of compression (recall the ice skater when she brings in her arms). For the same reason, it also must have an extremely powerful magnetic field. This means that as the star and its magnetic field spin around many times each second, energetic beams of radiation sweep around like a lighthouse beacon or searchlight. If the Earth happens to be in the direction of the beam, we see strong regular pulses on each sweep. Such objects are called pulsars, and they can be detected in radio waves, visible light or X-rays.

If in the aftermath of the supernova, the neutron star has a mass exceeding 3.0 times the mass of the sun, then nothing can prevent total ultimate collapse of the star to a point of zero radius and infinite density. Such an object is called a black hole, and it has such a strong gravitational field that not even light can escape from inside a boundary called the event horizon, which has a diameter of 4 miles times the mass of the collapsed remnant star. A 10 solar-mass remnant star would have an event horizon about 40 miles across. In the vicinity of a black hole, gravity becomes so strong that clocks near the hole would tick more slowly than clocks far away, according to Einstein's General Relativity. If we were to watch a spaceship fall into a black hole from somewhere outside its gravitational influence, we would see the spaceship's motion inward "freeze" at the event horizon because of the time distortion. Yet if we were on the falling spaceship, we would not notice any slowing of clocks as we passed through the event horizon, and within a fraction of a second, we on the ship would reach the singularity, the point of zero radius and infinite density. In all probability, we on the ship would be crushed and flattened instantaneously by enormous tidal forces.

Of course, single black holes cannot be detected because they give off no light! However, stellar black holes can be detected by their presence in a binary star system. A collapsed star in such a system reveals itself by the presence of hot X-ray emitting gas which it draws from its companion. Many such X-ray binaries contain neutron stars, but in at least a half-dozen binary systems, the mass of the collapsed companion, deduced from radial velocity changes in the spectral features of the primary star, exceeds 8 solar masses. In these systems, it is likely that the secondary companions are black holes.

Galaxies and Cosmology

In the 1850s, Lord Rosse observed many diffuse "spiral nebulae" with his large telescope; he had (as had Herschel before him) thought that they might be "island universes," distinct stellar systems separate from the Milky Way. Not until 1924 did Edwin Hubble establish that the "spiral nebulae" were actually other galaxies. He located Cepheid variable stars in the Andromeda nebula (also called Messier 31, or M31). These variables were observed to be near 18th magnitude, which from the Cepheid period-luminosity relation implies that M31 must be 2 million light years away, too remote to be within the Milky Way. Hubble and his colleagues photographed the spectra of many galaxies in the 1920s. To their astonishment, they discovered that with the exception of a few "local" galaxies like M31, every galaxy in the sky showed redshifted spectral lines, as if every gal-

axy in the sky were moving away from the Milky Way. Furthermore, they discovered that the farther away a galaxy is, the greater the redshift of a galaxy's spectral lines, and the faster it must be moving away from us. This famous relation is called Hubble's law of recession. The best available data indicates that for each megaparsec (1 million parsecs or 3.26 million light years) of distance between us and a galaxy, the speed of recession increases by about 67 kilometers per second (or 42 miles per second). We can use Hubble's law to determine distances to remote galaxies for which redshifts are measured. Currently, with the Hubble Space Telescope, we have detected distant galaxies as far away as 14 billion light years. The Universe is apparently filled with galaxies and groups of galaxies for as far as can be detected. Moreover, all these objects are moving away from us. How does one explain this observation?

The simplest interpretation of the law of recession is that the entire Universe is expanding. As it does so, space stretches like the skin of a balloon being inflated, or the dough in a loaf of raisin-bread rising in the oven. The galaxies are like dots on the balloon skin, or like raisins in the bread. As the balloon inflates or the dough rises, the dots or the raisins separate from each other. In the same way, as space expands, the galaxies separate, so that viewed from any one galaxy, all other galaxies seem to be moving away from it. Thus, our Milky Way is not the center of the Universe; in fact, because space is unbounded in this model, an observer can travel all over the Universe and not ever find a center or an edge. This is analogous to a microscopic creature restricted to the surface of an inflating balloon skin; it can move anywhere on the balloon skin, but will be unable to find a "center" or "edge."

Now the recession of galaxies can be further interpreted to mean that in the remote past, any two points in space were arbitrarily close together. Thus, we can project the expansion backward to a point of essentially zero radius and infinite density. In this view, the Universe begins with an explosive "Big Bang." If the expansion rate of the Universe has not changed since the Big Bang, then we can determine the age of the Universe by using the Hubble expansion rate (H_0) of 67 kilometers per second per million parsecs. We simply divide H_0 into 1000 and obtain a result of 15 billion years. A slightly faster expansion rate of 75 for H_0 implies a somewhat younger Universe of 13.3 billion years (shorter time for galaxies to separate to their present distances), while a slightly slower expansion rate of 60 for H_0 implies a somewhat older Universe of 16.7 billion years.

However, these calculations assume that the expansion of space is unaffected by the matter within it. One would expect that the gravitational attraction of galaxies towards each other would slow down the expansion of space over time, so that the expansion rate must have been higher in the past. This implies that the Universe is actually younger (by a factor of 1.5, or 9 to 11 billion years) than the 13.3 to 16.7 billion year range suggested by the current data on H_0. Unfortunately, this would mean that there are globular clusters in the Milky Way that are as old, if not older, than the Universe itself. Measurements of distant supernova explosions in 1999 have apparently resolved this dilemma; the observation that these supernovae are dimmer than expected (but otherwise identical to nearby stellar explosions in every other way) suggests that the expansion of space is accelerating over time, not slowing down as thought previously. This means that the Universe has likely been around for about 15 billion years, and there is no conflict of this with the ages of globular clusters.

Aside from the expansion of the Universe, what other evidence exists that the Universe had a beginning, which we refer to as the Big Bang? The most important piece of evidence in this regard comes from the discovery of the Cosmic Microwave Background Radiation (CMBR for short) in 1965. This shortwave radio noise has recently been observed to be the same intensity in every direction to 1 part in 100,000 by satellites in Earth orbit. This microwave radiation is from the Universe itself (which surrounds us). As we look out in space, we look back in time. If we look out 15 billion light years, we are seeing radiation that was emitted 15 billion years ago. The CMBR is radiation from that time. As the Universe evolves, space expands, and cools. Radiation travelling over billions of light years of expanding space will not slow down (the speed of light is constant), but will lose energy as the wavelength of light is stretched. After 15 billion light years, energetic visible photons (packets of light waves) will become microwave photons. Thus, the existence of the CMBR shows us that there was a hot dense phase of the Universe which no longer exists. Calculations indicate that since the end of this hot dense phase, the temperature of the expanding Universe should have dropped to about 3 degrees absolute, in almost precise agreement with the 2.7 degree microwave background actually measured.

It is of course clear that like the Universe itself, humanity's knowledge of the Universe continues to expand. It has been observed for a few decades within our Milky Way that the rotation speeds of stars about the galactic center show no decrease in speed even well beyond the Sun's orbit (unlike the planets in our Solar System). This means that much non-luminous mass exists even outside of the distance where stars are no longer seen, since otherwise the luminous mass would not be rotating nearly so fast. This so-called "dark matter" also exists in other galaxies, which show a similar constant rotation speed out to the edge of where the luminous matter ends. The "dark matter" also apparently exists in

between groups of galaxies. About 90% of the mass of spiral galaxies like the Milky Way is non-luminous, according to observations. What is this dark matter? To date, we are not certain, but cosmologists are very certain that the great majority of it is not baryonic matter, which is made up mostly of protons and neutrons, like stars, planets, and life. From an analysis of the CMBR by microwave detectors in space, combined with the analysis of distant supernovae discussed above, it now appears that about 30% of the mass in the known Universe is non-baryonic matter, while only about 5% is baryonic matter. Thus 95% of the stuff in the Universe is physically unobservable with our light-sensitive detectors! And what makes up the remaining 65%? The answer currently appears to be "dark energy," or quintessence. The dark energy, the energy contained within space itself, is apparently responsible for the acceleration of the Universe over time. Clearly our understanding of this new frontier is expanding as well.

The ancients thought that the Earth was the main object at the center of the Universe. The stars were lanterns hung up in the sky; the Sun was a chariot of fire, driven across the sky each day. Then Copernicus and Galileo showed that the Earth is but one of the systems of planets which goes around the Sun, and that the stars are other suns, very, very far away. Next it was suggested that the Universe consisted of the discus-shaped galaxy, its shape made apparent by the Milky Way, and that the Sun was at the center of the Milky Way. Then Harlow Shapley laid that idea to rest, demonstrating that the Sun is not at the center of the Milky Way. Only seven years later, Edwin Hubble determined that the "spiral or disc-shaped nebulae" were actually other "island universes" like the Milky Way. Now, astronomers, using large telescopes on Earth and in space, have discovered that there are billions of such galaxies scattered through space. The galaxies are hundreds of thousands of light years in diameter and millions of light years apart, extending out into space for billions of light years, and the end is not yet in sight.

Coincidentally, the four forces of nature (strong and weak nuclear forces, electromagnetism, and gravity, in order of decreasing relative strength) are such that if they were only a fraction different in relative strength, no life would exist in our Universe. If the Universe is an accident, the odds against it containing any order are ridiculously small, which of course suggests that it was not an accident. On the other hand, perhaps there are billions of universes, and we just happen to be in the one where the forces of nature have exactly the correct relative strengths for life to exist. Why does an entire Universe of galaxies exist then when all that is needed for life is one? This conundrum was perhaps best expressed by Albert Einstein: "The most incomprehensible thing about the Universe is that it is comprehensible."

Chapter 6
When the Sun Casts No Shadow

During the last half of May and again in mid-July each year, an event occurs in the Hawaiian skies that cannot take place anywhere in the mainland United States. The Sun passes directly overhead, and for an instant a slender, vertical object, such as a flagpole, will cast no shadow. This is possible because Hawai'i, located about 20 degrees North of the equator, is south of the Tropic of Cancer, at 23.5 degrees North, so that during the Sun's seasonal motion up to the Tropic of Cancer and back south again, it will at some point be directly above the latitude of Hawai'i. This phenomenon is popularly known in Hawai'i as "Lahaina Noon."

It is not hard to predict just when and where the Sun will be exactly overhead, with the help of The Astronomical Almanac. The chart below shows the paths which the Sun takes across the Hawaiian Islands from day to day during this period in May and early June. This is not the same every year because there is not an integer number of days in the year. Every four years, a correction is made with the addition of a day on February 29, the Leap Year correction. The lines shown here are correct for the years 2015, 2019 and 2023. The lines must be moved upward by 1/3 of the spacing between days, for the years 2018, 2022, and 2026. The lines must be moved upward

by 2/3 of the spacing between days for the years 2017, 2021 and 2025. Finally, they must be moved upward by one full day for the Leap Years 2016, 2020 and 2024. In other words, the line marked May 25 on the diagram will become May 24 in 2016. As another example, the line marked May 19 on the diagram passes right through Hilo in the year 2018, and again in 2022.

At the top of the chart is a series of numbers which indicate the time in minutes after 12 Noon Hawai'i Standard Time (HST) that the Mean Sun crosses our local meridian (a north-south circle that passes overhead) at its highest altitude. By definition, the Mean Sun is directly over the 150th meridian (150 degrees west longitude, some 540 miles east of Honolulu) at 12 Noon HST, but by the time it is over Hilo, near the 155th meridian, it is about 20 minutes after 12 Noon HST. Thus, the times at the top of the chart depend only on the longitude of the place of concern. Here are a few special cases of the local time correction that are useful:

Hilo:	20 minutes, 20 seconds
Maunakea:	21 minutes, 53 seconds
Haleakalā:	25 minutes, 0 seconds
Honolulu:	31 minutes, 29 seconds
Lihue:	37 minutes, 28 seconds

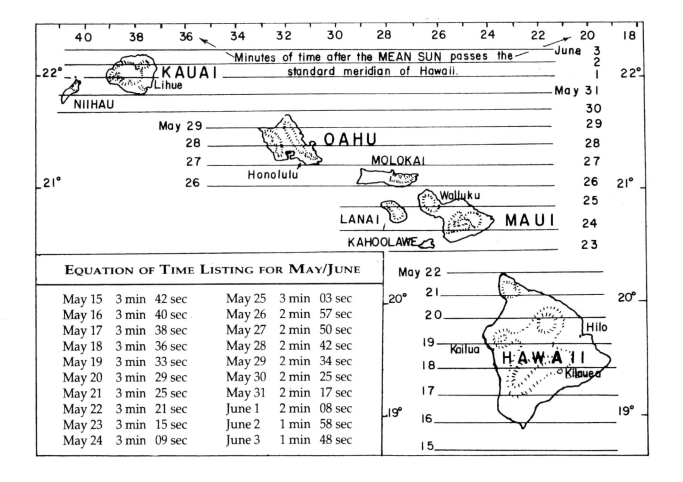

EQUATION OF TIME LISTING FOR MAY/JUNE

May 15	3 min	42 sec	May 25	3 min	03 sec
May 16	3 min	40 sec	May 26	2 min	57 sec
May 17	3 min	38 sec	May 27	2 min	50 sec
May 18	3 min	36 sec	May 28	2 min	42 sec
May 19	3 min	33 sec	May 29	2 min	34 sec
May 20	3 min	29 sec	May 30	2 min	25 sec
May 21	3 min	25 sec	May 31	2 min	17 sec
May 22	3 min	21 sec	June 1	2 min	08 sec
May 23	3 min	15 sec	June 2	1 min	58 sec
May 24	3 min	09 sec	June 3	1 min	48 sec

The Mean Sun (MS), however, is not the Sun we see; rather, the MS is the one that astronomers have invented so that they do not have to keep changing the clocks constantly. To find the time when the Sun is actually overhead, another correction must be made. The real or Apparent Sun (AS) may actually be ahead of or behind the imaginary or Mean Sun (MS) by an amount, measured in minutes, called the Equation of Time (ET). The Figure below shows how the ET varies throughout the year. When the ET is positive, the AS is ahead of the MS. When the ET is negative, the AS is lagging behind the MS. During May and early June, the ET is positive and so the Apparent Sun is ahead of the Mean Sun, which means it is seen on the meridian at its highest altitude before the Mean Sun.

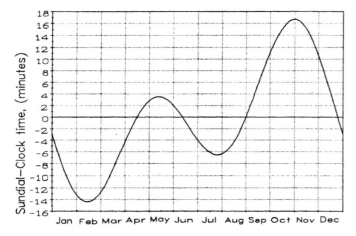

For example, on May 19, 2015, the ET is 3 minutes, 33 seconds, which means the Apparent Sun crosses the meridian of Hilo 3 minutes and 33 seconds before the Mean Sun, which in turn crosses the meridian 20 minutes, 20 seconds after 12 Noon HST. Thus, the Apparent Sun passes overhead earlier than the Mean Sun, giving the time of no shadow at 12:20:20 minus 00:03:33, or 12:16:47 HST (that is, 16 minutes, 47 seconds after 12 Noon HST).

The following table gives the difference for each day between May 15th and June 3rd. At this time of year the Apparent Sun is ahead of the Mean Sun. If you want to know when, in terms of HST, the real Sun will be on your local meridian (as shown on the scale on the top of the chart), you take the number of minutes that you are west of the standard meridian and subtract from it the Equation of Time for that date. That gives you the number of minutes after 12 noon HST that the real Sun is on your meridian.

Equation of Time listing for May/June

May 15	3 min	42 sec	May 25	3 min	03 sec
May 16	3 min	40 sec	May 26	2 min	57 sec
May 17	3 min	38 sec	May 27	2 min	50 sec
May 18	3 min	36 sec	May 28	2 min	42 sec
May 19	3 min	33 sec	May 29	2 min	34 sec
May 20	3 min	29 sec	May 30	2 min	25 sec
May 21	3 min	25 sec	May 31	2 min	17 sec
May 22	3 min	21 sec	June 1	2 min	08 sec
May 23	3 min	15 sec	June 2	1 min	58 sec
May 24	3 min	09 sec	June 3	1 min	48 sec

If you want to figure out when and where the Sun will be directly overhead in July, do just as we did. Hunt up a copy of The Astronomical Almanac for the current year — there are copies at the University of Hawai'i Library and libraries of the various observatory facilities in Manoa and Hilo. In Part C, find the section "Ecliptic and Equatorial Coordinates of the Sun — Daily Ephemeris." On the far left of the left-hand page, locate the date of interest. Look across to the column labeled "Apparent Declination" to find the latitude of the Sun for that day, and select the day when the Apparent Declination of the Sun is the same as your latitude. Then look all the way across to the right-hand column on the right-hand page to the column labeled "Ephemeris Transit." Subtracting this number from 12 gives the Equation of Time for that day. Now, both of these numbers are for 0 hours UT or Greenwich Mean Solar Time. By the time the Sun gets to Hawai'i, these numbers have changed towards the next day's values, so to get the correct values for Hawai'i you need to interpolate the difference between the two days. Since Hawai'i is about 10.5 hours from Greenwich, the interpolation factor is 10.5/24ths of the difference between one day and the next. The Equation of Time needs to be interpolated only if you want to get it closer than the nearest minute. In case you do not have access to a copy of this valuable reference book on the heavenly bodies, the following are approximate dates when the Sun will be overhead during its journey southward in July:

Over Kaua'i:	about July 11 to 14.
Over O'ahu:	about July 15 to 18.
Over Moloka'i:	about July 19.
Over Maui:	about July 20 to 22.
Over Hawai'i:	about July 23 to 29.

The Equation of Time during this period is minus 4 minutes to minus 6 minutes. In May the Apparent Sun is ahead of the Mean Sun. In July the Apparent Sun is behind the Mean Sun.

Chapter 7
The Calendar

Many millenia ago, early humans structured their lives around the movements of the celestial bodies in the sky above them. The Sun defined the day. So long as it was daylight they could roam about in search of food. At night they observed the phases of the moon and the manner in which that body moves around the Earth, or rather, to their perception, across the sky. These phases and movements became another measure of time; thus, the month was born.

The seasons came and went: the heat of summer, the changing color of the leaves in autumn, the cold of winter, the fragrant blossoms which burst forth in spring. Humans were keenly aware of the seasons, particularly when they started to harvest or raise crops. They watched the seasonal movement of the Sun north and south in its daily swing across the heavens. It reached a certain maximum altitude, which they were able to measure by the length of shadows, such as those of the pyramids. The length of the shadow foretold the flooding of the rivers that watered their crops; recurring events could be predicted. The year became a reality; the return of familiar stars to the same part of the sky could be measured in days.

Thus, the day, the month, and the year became units in the human calendar. It is unfortunate that these three units do not fit evenly into each other. The tropical year, within which the seasons occur, is 365.242198+ days long. The lunar or synodical month (new moon to new moon) has a length of 29.53058+ days.

Among many cultures over time the calendar has been based upon phases of the moon. These were the easiest to observe. The year was of secondary importance, perhaps because it was so long. When the growing of crops became an important part of life, as it did 10,000 or more years ago across the globe, the seasons also had to be brought into the calendar. This gave rise to the tropical year.

Julius Caesar, with the help of a distinguished astronomer of Alexandria named Sosigenes, revised the Roman calendar in 45 B.C. They divided the tropical year into 12 months (which did not attempt to agree with the moon phases). They also tried to solve the problem of what to do with the part of a day left over, by adding one day every fourth year. It was not immediately apparent that this left an error of three days every 400 years, but the error accumulated and threw off all the feast days. By 1582, when Pope Gregory revised the calendar again, they were off ten days. He had the ten days dropped from the calendar, much to the consternation and alarm of the people. England and her colonies in America did not get around to accepting this revision until 1752, and then eleven days had to be dropped. Russia held out against the change until 1918, and many nations clung to the Julian calendar until 1923. Pope Gregory's plan for correcting the difference was to omit leap year days, with their extra day, in all future centennial years (those ending in two zeroes) except when the number was divisible by 400. Dropping three leap years every 400 years reduces the error to about half a minute a year or about one day in 3,300 years.

Whereas most people living on the US mainland experience tremendous variations in temperature and weather marking the seasons of spring, summer, winter and fall, such seasonal changes are much more subtle in Hawai'i. Instead of the traditional four seasons, there are two distinct "seasons" in Hawai'i. The dry season with little rain is called Kau. The rainy season is called Ho'oilo and starts around November with the appearance of Makali'i (the Pleiades star cluster) rising at sunset. During this time, called Makahiki, little work could be done in the fields nor could war be waged because of the wet and muddy conditions, so it was designated as a time set aside for spiritual cleansing, dancing, sport, feasting, socializing, and tax collecting. This is also the season the koholā (humpback whales) come to Hawaii annually to give birth to their young.

Traditionally, Hawaiians lived in harmony with the changing appearance of the Moon (mahina) and used it to mark the passage of time. With so many tasks to be done—fishing, planting, warring, worshiping—the Moon's changing shape provided structure and pace to daily living. For example, fishing for certain types of fish or in particular locations was only allowed during specific moon phases to prevent overfishing. Other phases signaled when high tides occur at sunrise, making canoe launching easier, or when fishing should pause for regular preventative maintenance such as repairing nets. The phases of the moon were so important that farmers determined the best time to plant certain crops around the moon phases as well as the seasons.

As shown on page 57, the moon cycle was divided into three 10-night periods known as anahulu. The first was called ho'onui, "growing bigger," and began on the night of Hilo, the first crescent moon. The second anahulu was called poepoe, "round" or "full." The nights of the full moon – Akua, Hoku, and Māhealani – were known as na po mahina konane, meaning "bright moonlight." The last anahulu was called 'emi, "decreasing" or "waning." Muku, the new moon, marked the end of the cycle. Each night of the moon cycle was individually named and these names were memorized by all Hawaiian children and were a foundational part of Hawaiian life and education. A noted Hawaiian proverb alludes to a person who is ignorant as being like a child who has not yet learned the phases of the moon.

Chapter 8
Hawaiian Astronomy

The Hawaiians had an extensive knowledge of the heavenly bodies and of their apparent movements and they made considerable use of this knowledge in their daily life. They watched the movement of the moon and planets with reference to the position of the fixed stars and constellations. Hawaiian astronomers were called kilo hōkū, kilo meaning stargazer or observer, and hōkū meaning stars; therefore, kilo hōkū is one who observes and studies the stars.

"Every intelligent Polynesian had a clear idea of the cardinal points, north, south, east and west, and of the points midway," writes Dr. Kenneth P. Emory in his chapter on navigation in *Ancient Hawaiian Civilization*. There were names for the winds that came from these general directions and curiously, many of these wind and direction names are similar in different Polynesian groups. For example, in Hawai'i we have the word kona, meaning toward the southwest, and ko'olau, toward the northeast. In the area around Samoa their equivalents are Tonga and Tokelau, after which island groups to the south and north of Samoa have been named.

The Zenith, in Hawai'i, was called hikialoalo, and stars near the horizon were called hikianalia. The Milky Way, which stretches across the heavens, had various names, including Hōkūnohoaupuni, Paeloahiki, Kai'a and Leleaka, the first being the most common.

The Hawaiians distinguished readily between the planets, which they called hōkū'ae'a or hōkūhele, and the fixed stars, called hōkūpa'a. The planets were called by different names when they were in the eastern and western sky, just as we speak of "morning star" and "evening star." If not distinguished as a particular planet, the eastern morning star was Ho'omānalonalo (Venus or Jupiter). The evening star was Hōkūkomohana. Mercury was called Ukaliali'i (following the chief) because it was to be seen only close to the Sun; another name for it was Ka'āwela. Venus was called Hōkūao when in the eastern morning sky, and Hōkūkauahiahi when in the western evening sky. Mars, like other red objects in the sky, was called Hōkū'ula, or more specifically Holoholopīna'au, as well as 'Aukelenuiaiku. Jupiter was known as Ka'āwela, A'ohōkū (starlight), 'Iao (dawn), and also because of its brightness, Ikaika (strong, powerful). Saturn's name was Makulu (a drop of mist). The modern Hawaiian name for Uranus is Hele'ekela.

The Sun, known as Lā, was not worshipped in Hawai'i as in many regions, but it was regarded with great favor because of its usefulness in giving warmth, and in helping one to tell the time of day and direction. The Sun rose (hiki) in the direction kukula hikina, (east). The place where it set was komohana, (west). Facing the sunset, the right hand pointed toward 'akau (north) and the left hand to hema (south). An expanse or area could be indicated by using combinations of these direction words. The Hawaiian expression, "O Hawai'i ka la hiki, O Kaua'i ka la kau," indicated the expanse of the main islands of the Hawaiian chain: "Hawai'i is (in the direction of) the Sun arrived (east), Kaua'i is the Sun lodged (west)."

The Sun was the timekeeper of the day. There were names for its rising, its position half way up the eastern sky, on the meridian, turning over to go down the western sky and sinking into the west.

Various legends were told about the Sun. One of the most familiar of these is how it was snared by the demigod, Maui, to keep it from crossing the sky too quickly, in order to lengthen the day and allow time to dry the tapa made by his mother, the goddess, Hina. According to legend, Hina released the moon and stars from her calabash from which they flew up to take their places in the sky.

Another popular legend relates how Maui attempted to pull up a whole continent from beneath the ocean, but when his brothers, who were paddling the canoe, looked around and saw what he was doing, Maui's fishhook snapped and his efforts were prematurely terminated. Thus, the Hawaiian Islands were born. The fishhook flew up into the sky and became Kamakaunuiomaui, Maui's Fishhook, the constellation also known as Scorpio.

Hawaiian astronomical experts had the duty of announcing the correct time of year for preparing the soil, planting crops, harvesting, setting forth on ocean voyages, and even undertaking a battle with their enemies. From this it was but a step to call upon them to foretell or predict the outcome of all sorts of activities. This led people to regard them as akin to astrologers and oracles. Such a Hawaiian expert was called a kilo (seer, prophet, or judge; one who "looks earnestly"). Basically all this went back to a knowledge of the heavenly bodies and their apparent motions, upon which the Hawaiian calendar was based. The apparent movement of the stars across the sky, from east to west, both nightly and throughout the year, was quite familiar to the Hawaiians. They may not have understood that the nightly movement was due to the rotation of the earth on its axis and the yearly movement to its revolution around the Sun, but they made good use of these movements and could measure the time of night and the calendar of events with considerable accuracy.

The moon, mahina, is fundamental to the Hawaiian calendar. Each night of the month has a separate name that also refers to the following day. There are thirty such names, although some months only twenty-nine are used. The month begins with Hilo, the thin crescent moon that follows the first appearance of the new moon low in the western evening sky. That the calendar is very old in Polynesia is shown by the similarity between the names as used by the Hawaiians, Tahitians, Maori people of New Zealand, Rarotongans of the Cook Islands, Marquesans, and other Polynesian peoples. On the following page are those used by the Hawaiians, Maoris, Tahitians, Rarotongans, Marquesans, and Mangarevans.

In Hawai'i, four periods of kapu were observed each month during eight months of the year. The four months of the Makahiki period had no kapu periods. The Kapu of Ku began on the night of Hilo and was lifted the morning of Kūkahi. The Kapu of Hua began on the evening of Mōhalu, lasted two nights and a day, and was lifted on the morning of Hua. Hua means "an egg," and on the evening preceding its night and day the moon was slightly egg shaped, whereas on Akua, it was "distinctly round." The night and day of Hoku had two names: Hoku Palemo, if the moon set before daylight, and Hoku Ili, if the moon was still above the horizon when daylight came. At this point the astronomers knew whether there would be 29 or 30 days in the month. It may have been that adjustment was made at this point and the rest of the names were used each month. If the moon did not set until after sunrise the next day was called Māhealani; if rising was delayed until after darkness of night had set in, Kū Lua was used. The third kapu period was dedicated to Kanaloa, began on the evening of 'Ole Pau and ended the morning of Kaloa Ku Kahi. The Kapu of Kane began on the evening of Kane and was lifted the morning of Lono, when the moon rose at daybreak. Mauli found the moon "fainting," its rising delayed until daybreak had come. Muku found the moon "cut-off," when rising was delayed until the sun was so bright it could no longer be seen. In many parts of Polynesia - Hawai'i, Samoa, Tonga, Tahiti and the Marquesas, the new year began with the first new moon following the rising of the Pleiades in the eastern sky soon after sunset. At present this is in late November; 1000 years ago it would have been the first week in November; 2000 years ago about October 20. In the South Pacific it would have been a few days later. After the annual taxes were collected in Hawai'i, there was a period of festivities called Makahiki. The names of the months of the year vary in different localities, even in the same group of islands. One Hawaiian sequence is: Makali'i, Kā'elo, Kaulua, Nana, Welo, Iki'iki, Ka'aona, Hinaia'ele'ele, Hilina Ehu, Hilina Mā, 'Ikuwā, Welehu. It is not possible exactly to relate these with our calendar.

The Hawaiian astronomers were well aware that the month did not fit evenly into the year. At the end of the twelfth moon period there were ten or eleven days left over. We do not know exactly how the Hawaiian astronomers managed the details of this problem. Dr. Peter H. Buck (Te Rangi Hiroa) learned on the atoll of Manihiki, north of the Cook Islands, that they added a thirteenth month whenever it was needed. That is exactly what the ancient Greeks did. One of the Greek astronomers, named Meton, worked out a cycle of 12 years with 12 months and 7 with 13 months, repeating itself after the 19 years had gone by. The 3rd, 5th, 8th, 11th, 13th, 16th and 19th had 13 months. The Hawaiian kilo hōkū doubtless discovered this "Metonic cycle" in the same way Meton had -- by trying to make the month fit the year.

In summary, we know that the ancient Hawaiians' knowledge of astronomy allowed them to make use of the position and phases of the moon in reckoning time. They distinguished between the brighter planets and the "fixed stars" and had names for those visible to the naked eye. They recognized the return of stars to the same part of the sky after the interval of a year and kept track of seasons and the beginning of the year in this way. The Hawaiians were aware of the apparent movement of the sun north and south each year and were able to maintain courses at sea by following series of stars that rose or set at the same spot (pocket) on the horizon. Definite terms for a variety of astronomical concepts, such as zenith, horizon, major points of the compass, and groupings of stars, as well as several hundred individual stars were in common use. The kāhuna (ancient Hawaiian masters of the arts and sciences) held their knowledge closely and shared it only with students deemed worthy.

Many Hawaiian star names were recorded in the 19th century. Unfortunately, we cannot now identify some of these stars. The records do not associate the Hawaiian star names with their corresponding Latin or English names. Some of the principal Hawaiian star names in use today are identified with their corresponding western names, and shown on the star charts on pages 58 and 59.

The Hawaiian Moon Calendar
The Nights of the Moon

NIGHT	HAWAIIAN	MAORI	TAHITIAN	RAROTONGAN	MARQUESAN	MANGAREVAN
1.	Hilo	Whiro	Tirio, Teriere	'Iro	Ta-nui	Tu-nui
2.	Hoaka	Tirea	Hirohiti	Oata	Tu-hava	Hoata
3.	Kūkahi	Hoata	Hoata	Amiama	Hoata	Maheama-tahi
4.	Kūlua	Oue	Hami-ama-mua	Amiama-aka-oti	Mahea-ma-tahi	Maheama-rua
5.	Kūkolu	Okoro	Hami-ama-roto	Tamatea	Mahea-ma-vaveka	Maheama-toru
6.	Kūpau	Tamatea-akiri	Hami-ama-muri	Tamatea-aka-oti	Mahea-ma-hakapau	Maheama-riro
7.	'Olekūkahi (1st)	Tamatea-a-ngana	Ore-mua	Korekore	Koekoe-tahi	Korekore-tahi
8.	'Olekūlua	Tamatea-aio	Ore-muri	Korekore-aka-oti	Koekoe-vaveka	Korekore-rua
9.	'Olekūkolu	Tamatea-whakapau	Tamatea	O-Vari	Koekoe-hakapau	Korekore-toru
10.	'Olekūpau	Huna	Huna	'Una	Ai (Mahau)	Korekore-riro
11.	Huna	Ari-roa	Rapu, Ari	Maaru	Huna	Ari
12.	Mohalu	Mahwharu	Maharu	'Ua	Mahao	Huna
13.	Hua	Maurea	Hu'a	E-Atua	Hua ('Ua)	Maharu
14.	Akua	Atua-whakahaehae	Maitu	O-Tu	Atua	Hua
15.	Hoku (fullest)	Turu	Motu	Marangi	Hotu-nui	Etua
16.	Māhealani	Rakau-nui	Mara'i	Oruru	Hotu-maie	Hotu
17.	Kulu	Rakau-matohi	Tur'i, Turutea	Rakau	Tu'u	Maure
18.	Lā'aukūkahi	Tikirau	Ra'au-mua	Rakau-roto	Aniva (Akau)	Turu
19.	Lā'aukūlua	Oika	Ra'au-roto	Rakau-aka-oti	Matahi	Rakau
20.	Lā'aupau	Korekore	Ra'au-muri	Kore-kore	Kaau, Akau	Motohi
21.	'Olekūkahi	Korekore-tutua	Ore-mua	Korekore-roto	Koekoe-tahi	Korekore-tahi
22.	'Olekūlua (3rd)	Korekore-piri-ki-tangaroa	Ore-roto	Korekore-aka-oti	Koekoe-waena	Korekore-rua
23.	'Olekūpau	Tangaroa-a-mua	Ore-muri	Tangaroa	Koekoe-haapau	Korekore-toru
24.	Kāloakūkahi*	Tangaroa-a-roto	Ta'aroa-mua	Tangaroa-roto	Takaoa-tutahi	Korekore-riro
25.	Kāloakūlua*	Tangaroa-kiokio	Ta'aroa-roto	Tangaroa-aka-oti	Takaoa-vaveka	Vehi-tahi
26.	Kāloapau*	O-Tane	Ta'aroa-muri	O-Tane	Takaoa-hakapau	Vehi-rua
27.	Kāne	O-Rongo-nui	Tane	Rongo-nui	Puhiwa (Vehi)	Vehi-toru
28.	Lono	Mauri	Ro'o-nui	Mauri	Tane (Moui)	Vehi-riro
29.	Mauli	O-Mutu	Ro'o-mauri	O-Mutu	Nu-nui	Tane
30.	Muku (NEW)	Mutuwhenua	Mutu, Mauri-mate	Otire-o-avaiki	Nu-mata	Mouri

[*Kaloa is shortened from Ka'aloa.]

Hawaiian Moon Phases

Anahulu Ho'onui

| Hilo | Hoaka | Kūkahi | Kūlua | Kūkolu |

| Kūpau | 'Olekūkahi | 'Olekūlua | 'Olekūkolu | 'Olekūpau |

Anahulu Poepoe

| Huna | Mohalu | Hua | Akua | Hoku |

| Māhealani | Kulu | Lā'aukūkahi | Lā'aukūlua | Lā'aupau |

Anahulu Ho'ēmi

| 'Olekūkahi | 'Olekūlua | 'Olekūpau | Kāloakūkahi | Kāloakūlua |

| Kāloapau | Kāne | Lono | Mauli | Muku |

MAY
Star Chart 5 with Hawaiian names

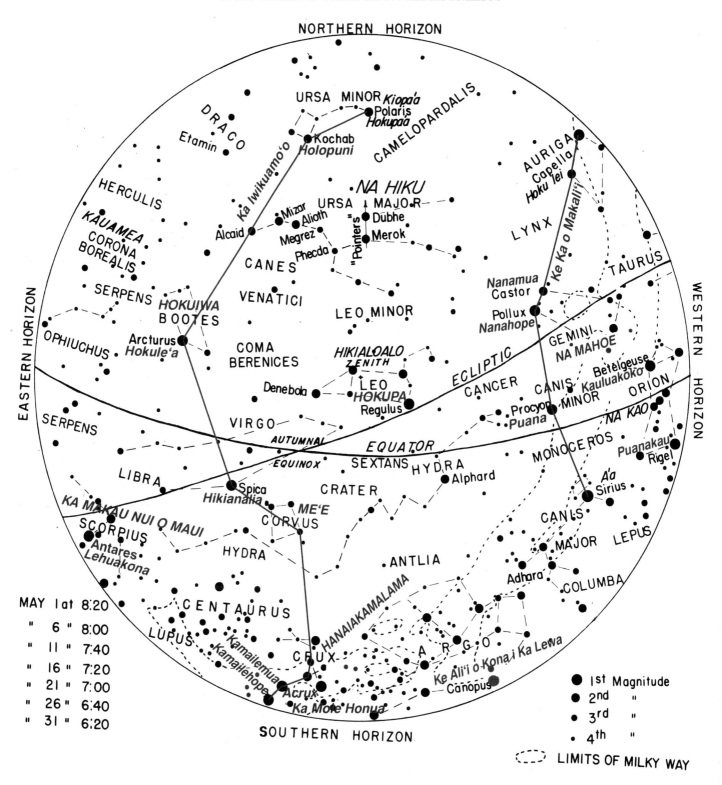

NORTHERN HORIZON

MAY 1 at 8:20
" 6 " 8:00
" 11 " 7:40
" 16 " 7:20
" 21 " 7:00
" 26 " 6:40
" 31 " 6:20

SOUTHERN HORIZON

● 1st Magnitude
● 2nd "
• 3rd "
· 4th "
(⌐ ⌐ ⌐) LIMITS OF MILKY WAY

NOTE:
To find the correct chart for viewing times and dates different from those listed please see Star Chart Finder and instructions on page 18. Constellations appear in capital letters. Stars appear in upper and lower case letters. Star lines are Ka Iwikuamo'o and Ke Ka o Makali'i. Hawaiian names are italicized.

NOVEMBER
Star Chart 11 with Hawaiian names

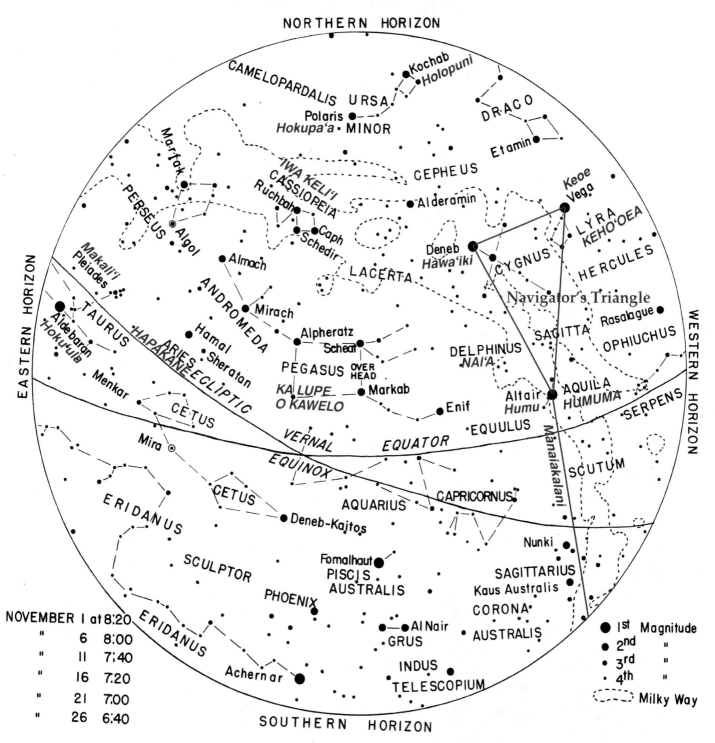

NOVEMBER I at 8:20
"   6   8:00
"   II   7:40
"   16   7:20
"   21   7:00
"   26   6:40

NOTE:
To find the correct chart for viewing times and dates different from those listed please see Star Chart Finder and instructions on page 18. Constellations appear in capital letters. Stars appear in upper and lower case letters. Star line is Manaiakalani. Hawaiian names are italicized.

Ka Heihei o nā Keiki

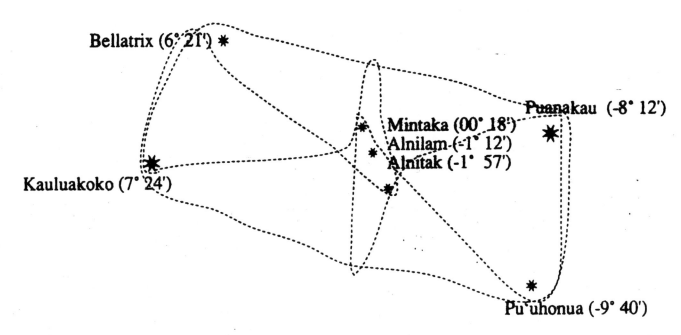

Bellatrix (6° 21')

Puanakau (-8° 12')

Mintaka (00° 18')
Alnilam (-1° 12')
Alnitak (-1° 57')

Kauluakoko (7° 24')

Pu'uhonua (-9° 40')

The constellation Ka Heihei o nā Keiki was named for its resemblance to a string figure (heihei) made by children. The string figure which the configuration of stars most closely resembles goes by the names of "hoku" (star), "spider," or "kohe ekemu" (embrace me), a continuation of the figure "po" (night) which represents a starry night and allows the player to make the stars appear, as in the evening, and disappear, as at dawn. Both figures, "hoku" and "po," were made on all the Hawaiian islands. "Spider" is an appropriate name since this figure travels along the celestial equator, which was called Ke Alanui o Ke Ku'uku'u, or "The Roadway of the Spider." Mintaka, one of the three middle stars in this constellation, with a declination of 0° 18', rises almost due east and sets almost due west since it is on the celestial equator.

See Lyle A. Dickey, *String Figures from Hawai'i*, (Honolulu: Bishop Museum, 1928, pp. 78-80), for instructions on how to make "po" and "hoku" or "spider," as well as the chants that go along with the two figures.

Chapter 9
Polynesian Voyaging and Wayfinding
"O na hoku no na kiu o ka lani". *The stars are the eyes of heaven.*
('Olelo No'eau 2513).

Sooner or later those interested in Hawaiian culture and voyaging will probably hear the story of the "sacred calabash." This is supposed to have been the sextant-like navigating instrument which the Polynesians are said to have had when they made their long voyages to Hawai'i in great double canoes. It is described as bowl-shaped or keg-shaped, with a series of holes around the upper part of the sides, equidistant below the rim. In order to make sure that it was held level, the calabash reportedly was filled with water up to the holes. When no water spilled out any hole, it was level. The holes were bored at such a distance from the rim that, in the latitude of Hawai'i, one could look through the hole on one side and just see the Pole Star over the opposite rim.

The story goes on to describe how the navigator sailed northward until he could just see the Pole Star in this fashion; then he would sail westward until the mountain peaks of Hawai'i came into view and would make his landfall. This instrument is so simple and its use so plausible that it catches the imagination and understanding of everyone who hears about it. It really is a pity that scientists who have investigated the subject are unable to substantiate the story. The Polynesians clearly navigated by means of the stars, but what is not so clear is the role of the "sacred calabash."

A principal source of information regarding this sacred calabash comes from the late Rear Admiral Hugh Rodman, USN. When he was a young lieutenant he visited Honolulu, was entertained at the Palace, and according to his story was shown this remarkable instrument and told its story by King Kalākaua.

The King had a large calabash with a very interesting history, which he kept in the Palace and showed visitors with pride. He had it decorated with a series of gold bands on which various historic scenes were engraved. The wooden cover had a gold plate that told its history. More than half a century ago, this "sacred calabash" was placed in Bernice P. Bishop Museum. The translation of the plaque on its cover is as follows: "The wind container of La'amaomao that was in the keeping of Hauna, personal attendant of Lonoikamakahiki I. It passed on to Paka'a, a personal attendant of Keawe-nui-a-'Umi. It was placed in the royal burial cave of Ho'aiku, on the sacred cliff of Keoua, at Ka'awaloa, island of Hawai'i.

Received by King Kalākaua I from Ka'apana, caretaker of Ho'aiku."

Admiral Rodman used a picture of this calabash to illustrate his story, so it is unlikely that he had any other calabash in mind. The question is, could this have been used as a navigating instrument and, if not, what was its use? This calabash has a series of holes, in threes, around the rim. Their distance from the edge is such that sighting through even the lowest of the three and over the opposite rim would give an angle of about 11 degrees. The calabash measures more than 33 inches high and a foot in diameter. Filled with water up to the level of the holes, it would weigh more than 100 pounds. With such a weight it would be very difficult to hold level at arms length in a bobbing canoe, let alone to use as a sextant. Could the "sacred calabash" have been a watertight traveling trunk, a container for such things as featherwork, fine tapa, ornaments and other possessions on a sea voyage or in the home? Most of them were made of large gourds covered with a plaited or twined reinforcement of fiber, the aerial rootlets of the 'ie'ie (Freycinetia arborea). A number of them are preserved in Bishop Museum and others are known. The series of holes was used in tying on the gourd or wooden cover, but in no case do the holes give an angle as large as 20 degrees, the latitude of Hawai'i.

The story of the calabash has been intertwined with that of the Hawaiian gourd compass, first related to Theodore Kelsey by David Malo Kupihea (see *Na Inoa Hoku*, 1975, by Rubellite K. Johnson and John K. Mahelona). A navigation gourd compass might be from about 1.5 to 3 feet in diameter, and at least 4 to 6 inches deep. It might be made of any suitable wood such as kou or milo. There were two sight holes; one was aligned with Polaris. At intervals of 40-45 degrees at the rim of the gourd were double hitches called pu'umana. Across the top of the gourd a net was placed; special terms were used to designate each mesh square (maka) or pu'umana knot ('alihi) at the circumference. Stars were reflected in the water and seen through the 36 mesh squares; as the voyage proceeded, the stars were tracked across the net. The star in the east was called the Hokuiwa, or frigate bird. Each of the nine principal guide stars was represented by one of the pu'umana knots around the edge of the gourd rim. In a single canoe, the gourd might be hung up on the mast, or fastened to the covering piece over the bow, by placing the net with meshes over it and tacking down the extending cords around it. The man in the bow would then be the kilo, or observer. According to Kupihea and Kelsey, navigation gourds (smaller versions of the calabash) were used on local trips to other islands, although this is difficult to substantiate.

If they did not use a calabash-sextant, how then did the Polynesians navigators observe the stars in navigating their double canoes across the Pacific? Observe the stars they did, and they very likely needed no mechanical sextant to do so. In addition, they had extensive knowledge of the winds, ocean swells and currents, the flight of birds and the subtle arrangement of clouds over islands. They had good judgment in estimating angles. Crossing the equator, we can watch the pole star come into view above the horizon, and climb slowly up the sky as we proceed northward. We would have no trouble judging the north latitude within one degree. A skilled Polynesian navigator certainly could have done better. Long before his canoe had come within sixty miles of an island, the flight of birds would have told him of the presence and direction of land.

Ancient Polynesian methods of navigation by means of stars, winds, swells, currents, cloud tints and the flight of birds, have been rediscovered in Hawai'i through the people of the Tuamotu Archipelago. They still use these methods, which were almost forgotten in most parts of Polynesia. Direction is kept at night by steering toward one after another of a definite series of stars, when these guiding stars are near the horizon. If heading in a westerly direction, these stars are those that seem to sink into the same "pocket," or "house," of the sky. Heading in an easterly direction, the guiding stars all rise out of the same "house." A voyage between distant islands generally is undertaken at a certain time of year, when winds, currents and conditions at sea are likely to be favorable. The list of guiding stars for each course and season was a definite part of the sailing directions. A favorite time to set out for islands to the south was autumn, and A'a (Sirius) was one of the guiding stars at that time.

Much of the star lore connected with voyages to Hawai'i has been lost. Even the modern equivalents of Hawaiian star names are gone, because the persons who recorded these names, which their Hawaiian informants recited, did not themselves know the English or Latin names of the stars and constellations. More than 200 names of heavenly bodies known to the Hawaiians are listed, but only a small part of them can be identified today. A principal reference for such information is *Na Inoa Hoku* (1975), by Rubellite K. Johnson and John K. Mahelona.

Early Hawaiian tradition is full of accounts of famous navigators. Some of these are noted by Bruce Cartwright in "Some Ali'i of the Migratory period," (Bishop Museum Occasional Papers, 10 (7), 1933); others in folklore were gathered by Judge Abraham Fornander and published in Bishop Museum Memoirs, volumes 4 to 6, and his *An Account of the Polynesian Race*, volume 1, London, 1878.

There is the story of Pā'ao, one of the pioneer settlers of Puna, Hawai'i. He had with him an astronomer and navigator as well as a sailing master. Finding the Hawaiian Islands a good place to live, he returned to Kahiki to get a chief who could rule over the little band of settlers in that part of Hawai'i. His return voyage from the Society Islands with the chief Pili may have helped to establish the system of sailing directions for this route.

Another great navigator, Kaulu, the son of Kalana, navigated from Hawai'i to the southern islands and brought back a famous priest, navigator and astronomer named Luhaukapawa. Still another famous navigator and astronomer of Hawaiian tradition was Kamahualele. He accompanied Mo'ikeha, a chief of O'ahu, on his voyage from Waipi'o, Hawai'i, to Tahiti and back to Kaua'i. He then returned to Tahiti with Kila, youngest son of Mo'ikeha and again returned safely to Kaua'i.

These were hardy men, who could stand long periods of exertion on limited supplies of food and water. They understood the sea and its currents and the winds. They were guided by the position of the Sun by day and of the stars at night. They made landfalls by watching the flight of birds. They believed that they were never separated from the gods and demi-gods of the ocean depths, and the materials of life. The seafaring Polynesians were the greatest ocean explorers of their time and with the use of the large double-hulled canoes (wa'a kaulua), they arrived from Nuku Hiva (Marquesas) and Tahiti. The canoe captain, or ho'okele, likely did not need a compass, chart or sextant to make the voyage.

Mau's Compass

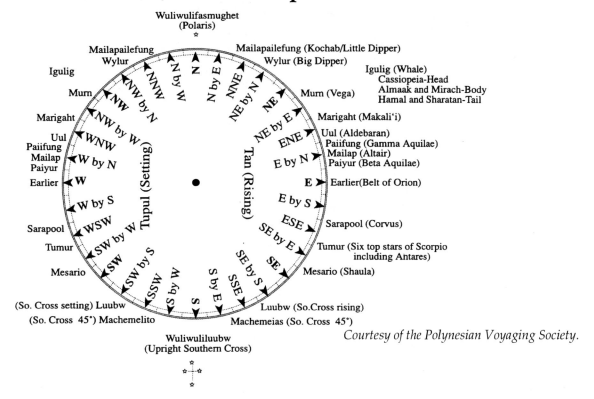

Courtesy of the Polynesian Voyaging Society.

By the time Europeans arrived in Hawai'i in the 18th century, voyaging between Hawai'i and the rest of Polynesia had ceased for more than 400 years. The reason for decline of voyaging is not known, but archaeological evidence suggests a dramatic expansion of population and food production in the Hawaiian Islands. Perhaps ties to families and gods in Polynesia weakened over time as the society in Hawai'i flourished

The Polynesian Voyaging Society was established in the early 1970s to rediscover the ancient art and science of Hawaiian navigation. The wa'a (canoe) Hokule'a was built between 1973-1975, and made its first historic voyage from Hawai'i to Tahiti, guided without instruments by traditional Micronesian navigator Mau Piailug. In 1980, Hawaiian Nainoa Thompson, a student of Mau's, successfully navigated Hokule'a to Tahiti and back to Hawai'i. In 1985-87, Hokule'a voyaged to Aotearoa (New Zealand) and back via Tahiti, the Cook Islands, Tonga, Samoa and Tuamoto; in 1992 she went to Tahiti and the Cook Islands, and in 1995, to Nuku Hiva (Marquesas). In 1999-2000, 25 years of voyaging achievement culminated with another historic voyage to Rapa Nui (Easter Island), the most isolated island in Polynesia, as well as to the Marquesas and Pitcairn Island.

The traditional navigation system used by Mau Piailug, illustrated above, designates 32 distinct houses around the horizon at unequal angular intervals. A star rises, like the Sun, in a particular house on the eastern horizon, travels across the sky, and sets in a correspond-ing house on the western horizon. On the Hawaiian Star Compass designed by Nainoa Thompson, page 64, the 32 houses are at equally spaced angular intervals. The house in which a star rises has the same name as the house in which it sets. The house that a star sets in is at the same angular distance and in the same direction from west as the house in which it rose in the east. Thus, the recognition of a rising or setting star and the knowledge of the house in which it rises and sets gives the observer a directional point of orientation. The rising points of the 21 brightest stars, the canoe-guiding stars, or Hōkūho'okelewa'a on the star compass are for stars rising at the Equator. As the observer moves away from the Equator, rising and setting points shift north for stars rising north of east, and south for stars rising south of east. At the north and south poles, of course, stars travel in circles at fixed altitudes around the sky, without rising or setting. The Moon (mahina) rises about 48 minutes later each night at a different position on the eastern horizon from the night before. Its rising point moves back and forth between ENE - 'Aina Ko'olau and ESE - 'Aina Malanai during its 29.5 day cycle. It sets between WNW - 'Aina Ho'olua and WSW - 'Aina Kona.

Nainoa Thompson used three major star groups to navigate; these three groups are Ke Ka o Makali'i (The Canoe-Bailer of Makali'i), Ka Iwikuamo'o (The Backbone), and Manaiakalani (The Chief's Fishline). A fourth star group has now been added, Ka Lupe o Kawelo (The Kite of Kawelo). Ke Ka o Makali'i is formed by five stars

Hawaiian Star Compass

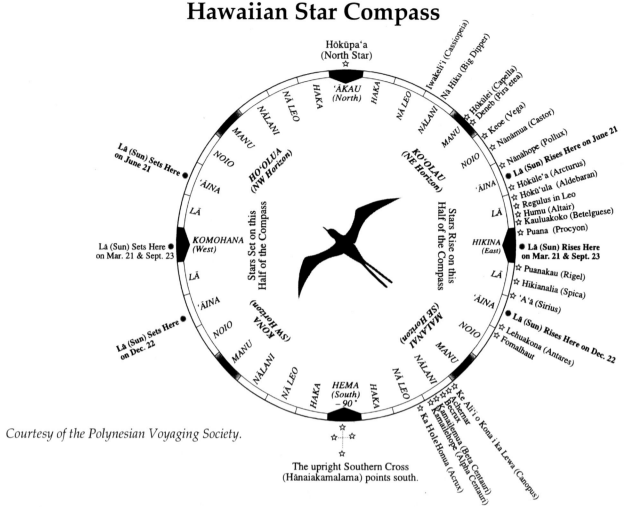

Courtesy of the Polynesian Voyaging Society.

The upright Southern Cross
(Hānaiakamalama) points south.

curving across the sky from north to south in the shape of a bailer; these five stars are Hokulei (Capella), Namahoe (Castor and Pollux), Puana (Procyon), and A'a (Sirius). Ka Iwikuamo'o runs from Hōkūpa'a (Polaris) at the north celestial pole to Hanaiakamalama (the Southern Cross) near the south celestial pole and is seen as vertebrae along a backbone. The star line includes Hōkūpa'a (Polaris), Holopuni (Kochab), Nahiku (The Big Dipper), Hōkūle'a (Arcturus), Hikianalia (Spica), Me'e (Corvus), Hanaiakamalama (the Southern Cross), and Nakuhikuhi (Alpha and Beta Centauri or "The Pointers"). Manaiakalani goes from 'Iwakeli'i (Cassiopeia) in the north to Kamakaunuiomaui (Scorpio, also called Maui's Fishhook) in the south, and is dominated by the Navigator's Triangle, comprised of Hawa'iki (Deneb), Keoe (Vega), and Humu (Altair). Kamakaunuiomaui is on the opposite side of the sky from Nakao (referring to the belt of three stars in Orion), or Kaheiheionakeiki, (referring to the whole Orion star group). The northern part of Kalupeokawelo is made up of 'Iwakeli'i (Cassiopeia) and Kalupe (the Great Square of Pegasus - the southern part is made up of the stars Fomalhaut, Alnair, Dipha, Anka'a, and Achernar).

After several decades of Hawaiian cultural and voyaging renaissance, first envisioned by the pioneering members of the Polynesian Voyaging Society, a Pacific-wide movement has reached fruition in the widespread revival of Polynesian voyaging traditions. Before his death in 2010, Mau Piailug designated five Hawaiians as Master Pwo Navigators charged with keeping the wayfinding traditions and navigational knowledge alive: Chad Kalepa Baybayan, Milton "Shorty" Bertelmann, Bruce Blankenfeld, Chadd 'Onohi Paishon, and Charles Nainoa Thompson. Together with apprentice navigators, the Hokule'a and its sister canoe Hikianalia embarked in 2014 on a 3-year, 85-port, 26-country, 48,000 mile circumnavigational voyage around Earth. The educational and diplomatic voyage is known as "Malama Honua," which means "to care for our Earth." While voyagers on Hokule'a will strictly use the stars, ocean current, winds, and birds as mapping points for navigation, the Hikianalia is outfitted with state of the art technology and satellite communication systems to allow crewmembers to communicate with classrooms, the media and the world via live chats, videos, blog posts, and photographs and to help keep the Hokule'a safe in the face of danger.

Hokule'a and Hikianalia marry ancient and modern technology to bring Polynesian voyaging traditions into modern times. Similarly, ancient Hawaiian astronomical knowledge is complemented by modern discoveries about the skies above the beautiful islands of Hawai'i.

Star Compass
Ke Ka o Makali'i ~ The Canoe-Bailer of Makali'i

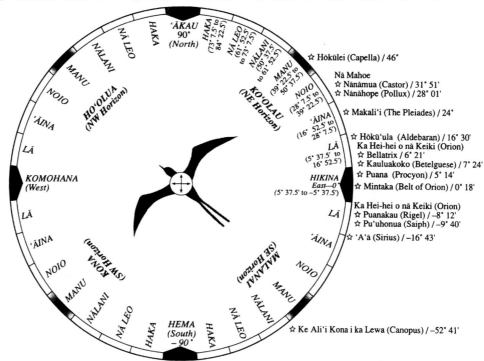

☆ Hōkūlei (Capella) / 46°

Nā Mahoe
☆ Nānāmua (Castor) / 31° 51'
☆ Nānāhope (Pollux) / 28° 01'

☆ Makali'i (The Pleiades) / 24°

☆ Hōkū'ula (Aldebaran) / 16° 30'
Ka Hei-hei o nā Keiki (Orion)
☆ Bellatrix / 6° 21'
☆ Kauluakoko (Betelguese) / 7° 24'
☆ Puana (Procyon) / 5° 14'
☆ Mintaka (Belt of Orion) / 0° 18'

Ka Hei-hei o nā Keiki (Orion)
☆ Puanakau (Rigel) / –8° 12'
☆ Pu'uhonua (Saiph) / –9° 40'

☆ 'A'ā (Sirius) / –16° 43'

☆ Ke Ali'i Kona i ka Lewa (Canopus) / –52° 41'

Star Compass
Ka Iwikuamo'o ~ The Backbone

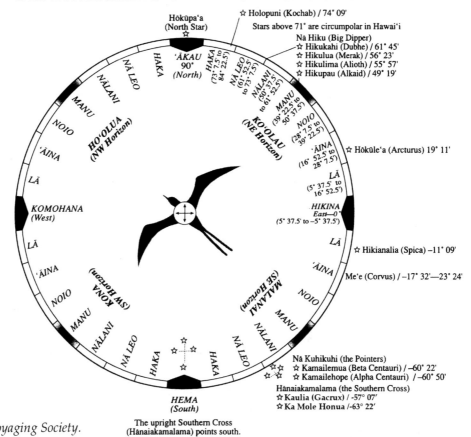

☆ Holopuni (Kochab) / 74° 09'

Stars above 71° are circumpolar in Hawai'i

Nā Hiku (Big Dipper)
☆ Hikukahi (Dubhe) / 61° 45'
☆ Hikulua (Merak) / 56° 23'
☆ Hikulima (Alioth) / 55° 57'
☆ Hikupau (Alkaid) / 49° 19'

☆ Hōkūle'a (Arcturus) 19° 11'

☆ Hikianalia (Spica) –11° 09'

Me'e (Corvus) / –17° 32'—23° 24'

Nā Kuhikuhi (the Pointers)
☆ Kamailemua (Beta Centauri) / –60° 22'
☆ Kamailehope (Alpha Centauri) / –60° 50'
Hānaiakamalama (the Southern Cross)
☆ Kaulia (Gacrux) / -57° 07'
☆ Ka Mole Honua / -63° 22'

The upright Southern Cross
(Hānaiakamalama) points south.

Courtesy of the Polynesian Voyaging Society.

Star Compass
Manaiakalani ~ The Chief's Fishline

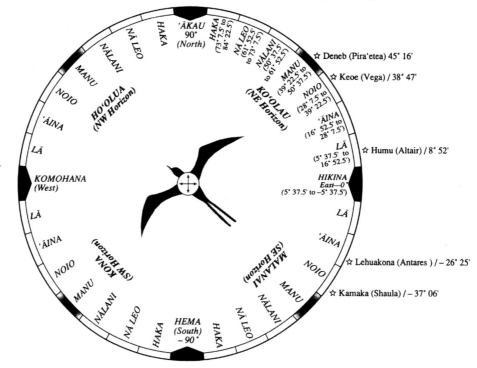

Star Compass
Ka Lupe o Kawelo ~ The Kite of Kawelo

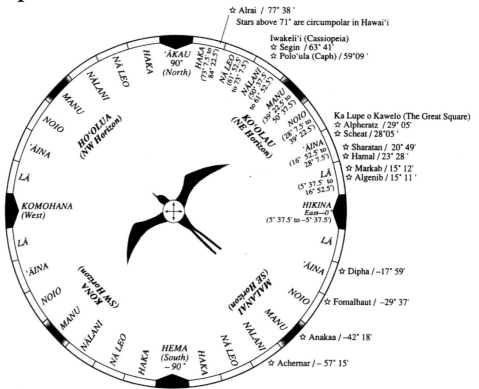

Courtesy of the Polynesian Voyaging Society.

Edwin H. Bryan, Jr.

The first edition of *Stars Over Hawai'i* was written by E.H. Bryan, Jr., in 1955. A revised edition was published in 1977. Since his death in 1985, no further revisions have appeared until now when it is deemed appropriate and worthwhile to update this very popular handbook of astronomy in Hawai'i.

Ed Bryan, as his friends knew him, was a fount of knowledge in all aspects of the natural history of Hawai'i. He was born in California in 1898 but came to Hawai'i at the young age of 18. He studied at the University of Hawai'i where he earned a Masters degree in entomology and carried out additional graduate studies at Stanford University in botany and zoology.

He was one of the early users of the Kaimuki Observatory which was built in 1910 primarily to observe Comet Halley's 1910 apparition. During 1917 and 1918, Bryan and fellow amateur R. W. French, a sergeant in the U.S. Army Medical Corps, used the telescope in the observation of variable stars. Both were members of the American Association of Variable Star Observers. Bryan also helped organize and conduct public star parties at the Kaimuki Observatory. (As a footnote, the Kaimuki Observatory, once located on Sea View Avenue in Kaimuki, was demolished in 1958.)

During the 1930's and 1940's a growing number of amateur astronomers felt the need for some sort of organization and the availability of astronomical information tailored to Hawai'i. Bryan, now on the staff of the Bernice P. Bishop Museum, responded to this latter need by preparing the original *Stars Over Hawai'i*. Most importantly, it contained a star chart for each month of the year, for the latitude of Hawai'i. This book received wide circulation and certainly must have had a significant impact on astronomical literacy in Hawai'i. Bryan also initiated the monthly publication in a local newspaper of the current star chart and a description of astronomical phenomena for the month, a tradition that has continued uninterrupted to the present day.

Ed Bryan also authored countless popular articles on astronomy and natural history in Hawai'i. When I initially came to Hawai'i in 1945, the first book I found to teach me about beautiful Hawai'i was Bryan's little volume, *Hawaiian Nature Notes*, first published in 1933. It is my opinion that E. H. Bryan, Jr., more than any other individual, served to inform and stimulate public interest in Hawaiian astronomy and natural history during the early decades of the 20th century.

Walter R. Steiger, Ph.D.
January 2002

Dr. Richard A. Crowe

The late Dr. Richard Crowe, Professor of Astronomy at the University of Hawai'i at Hilo (UHH), and Astronomer-in-Residence at the 'Imiloa Astronomy Center of Hawai'i from 2006 to 2012, was born in Canada. He obtained his B.Sc. and M.Sc. in astronomy from the University of Western Ontario and his Ph.D. from the University of Toronto. From 1977-79, he worked as the Resident Observer for the University of Toronto Southern Observatory at Las Campanas, Chile. Following the completion of his Ph.D. in 1984, he moved to the island of Hawai'i working for three years as the Canadian Resident Astronomer for the Canada-France-Hawaii Telescope (CFHT) Corporation based in Kamuela.

After coming to UHH in 1987, his teaching responsibilities ranged from introductory physics and astronomy to senior level astrophysics and quantum mechanics; he was recognized for his teaching in 2005 (AstroDay Award) and 2010 (UHH Taniguchi Award). Dr. Crowe's main research interests and body of published work were in the areas of pulsating stars, stellar evolution and spectroscopy. He also published a dozen scholarly articles about science education and criticism of pseudoscience. In 1991, Dr. Crowe was selected as a Fujio Matsuda Fellow by the Matsuda Scholars and Fellows Selection Committee. He was Chair of the UHH Department of Physics and Astronomy (1992-2002), and the Principal Investigator on the New Opportunities through Minority Initiatives in Space Science (NOMISS) program funded by NASA's Office of Space Science (2001-2004). In 2001, he and Alice Kawakami won City Bank's TIGR Award in Astronomy for their NOMISS community outreach efforts. The NOMISS program was designed to encourage local and Hawaiian students to enter careers in space science, by working with K-12 teachers and undergraduates to integrate astronomy with Polynesian skylore, voyaging, and Hawaiian culture.

Dr. Crowe remained an active participant and coordinator with the Journey Through the Universe flagship outreach program, pioneered in East Hawai'i, until his untimely passing in May of 2012. The Richard Crowe Memorial Scholarship has been established at the University of Hawaii at Hilo to benefit deserving students of Astronomy and Physics. Information regarding donations can be found on the copyright page.

Dr. Walter Steiger
Professor Emeritus of Physics and Astronomy
University of Hawai'i at Mānoa

The late Dr. Walter R. Steiger came to UH Mānoa in 1953 as a graduate student in Physics. After receiving his Ph.D. from the University of Cincinnati, he returned to UH Mānoa in 1958 as the third member of the Physics Department. During his career at Mānoa, he began an astronomy program and developed a solar observatory at Makapu'u Point on O'ahu, that later was supplanted by one on Haleakalā, Maui. Dr. Steiger served two years as Provost at Kaua'i Community College, and served on the UH Board of Regents for four years. After an early retirement in 1980, he entered a new form of teaching as the manager of the Science Center at the Bishop Museum Planetarium. It was during this period that he had the good fortune to work with Ed Bryan, author of this book. In 1987, looking for another new challenge, he was offered the opportunity to work on Mauna Kea as Site Manager for the Caltech Submillimeter Observatory. After retirement from CSO, he was back to his first love, teaching physics, this time at UH Hilo, as a part-time lecturer. Until his passing in February 2011 he spoke and wrote about the origins of astronomy in Hawai'i and helped CSO with its outreach program. Dr. Steiger's association with the 'Imiloa Astronomy Center reflected his enthusiasm for bringing the science of astronomy to the general public. He was honored for his service there with a special ceremony in October 2010. Dr. Steiger is considered a person who was instrumental in establishing modern astronomy in Hawai'i.

Dr. Timothy F. Slater
Professor and Endowed Chair of Higher Education
University of Wyoming at Laramie

Dr. Tim Slater is an Astronomer at the University of Wyoming where he holds the Wyoming Excellence in Higher Education Endowed Chair for Science Education. He is the Editor-in-Chief for the Journal of Astronomy & Earth Sciences Education and a Senior Scientist at the CAPER Center for Astronomy & Physics Education Research. A frequent lecturer on contemporary strategies for innovative astronomy science teaching, Professor Slater is often introduced as the Professors' Professor because of his decades-spanning work on teaching professors how to teach. He has worked closely with astronomers throughout Hawai'i on helping them improve the effectiveness of their visits to local K-12 classrooms. Over the last several decades, Professor Slater has worked with hundreds of Hawaii's K-12 teachers on implementing best strategies for teaching astronomy. He lives part of the year in Hawai'i and part of the year in Wyoming. Professor Slater earned his Ph.D. at the University of South Carolina, his M.S. from Clemson University, and two bachelors' degrees from Kansas State University. Dr. Slater has been elected for multiple-terms on the Councils and Boards of Directors for the American Astronomical Society, Astronomical Society of the Pacific, National Science Teachers Association, Society of College Science Teachers, and has chaired committees for the American Association of Physics Teachers, American Physical Society, and American Institute of Physics. He is an author on more than 100 refereed articles, eleven books, winner of numerous awards, and is frequently an invited speaker on improving teaching of astronomy.

Resources

ASTRONOMY

Caltech Submillimeter Observatory (CSO)
CSO, located on Maunakea, is one of the world's premier facilities for astronomical research and instrumentation development at submillimeter wavelengths.
www.cso.caltech.edu/cso.html

Canada-France-Hawaii (CFHT)
CFHT is a versatile and state-of-the-art astronomical observing facility which is well matched to the scientific goals of it's user's community and which fully utilizes the potential of the Maunakea site.
www.cfht.hawaii.edu

Gemini
The Gemini Observatory is a multi-national partnership consisting of twin 8.1-meter diameter optical/infrared telescopes located on two of the best observing sites on the planet, mountains in Hawai'i and Chile. Gemini Observatory's telescopes can collectively access the entire sky.
www.gemini.edu

International Astronomical Union (IAU)
The mission of IAU is to promote and safeguard the science of astronomy in all its aspects through international cooperation.
www.iau.org

James Clerk Maxwell Telescope (JCMT)
With a diameter of 15m the JCMT is used to study our Solar System, interstellar and circumstellar dust and gas, and distant galaxies.
www.eaobservatory.org/jcmt/about-jcmt

NASA- National Aeronautics and Space Administration
"To reach for new heights and reveal the unknown so that what we do and learn will benefit all humankind"
www.NASA.gov

Smithsonian Astronomical Submilimeter Array Telescope (SMA)
SMA explores the universe by detecting light of colors which are not visible to the human eye. It receives millimeter and submillimeter radiation.
www.cfa.harvard.edu/sma/general

Subaru Telescope
Subaru Telescope is one of the world's largest and most technologically advanced telescopes. Through the open use program astronomers throughout the world have access to Subaru's excellent image quality.
www.subarutelescope.org

Thirty Meter Telescope / TMT
TMT will enable astronomers to study objects in our own solar system and stars throughout our Milky Way and its neighboring galaxies, and galaxies at the very edge of the observable Universe, near the beginning of time.
www.tmt.org

United Kingdom Infared Telescope/ UKIRT
UKIRT is supported by NASA and operated by the University of Hawai'i, the University of Arizona, and Lockheed Martin Advanced Technology Center; operations are enabled through the cooperation of the Joint Astronomy Centre of the Science and Technology Facilities Council of the U.K.
www.ukirt.hawaii.edu

W. M. Keck Observatory
Keck Observatory operates the two largest and most scientifically productive telescopes on Earth, opening the Universe to the world, from the summit on Maunakea in Hawai'i.
www.keckobservatory.org

EDUCATION AND THE ARTS

Bernice Pauahi Bishop Museum
The Bishop Museum's mission is to be a gathering place and educational center that actively engages people in the presentation, exploration and preservation of Hawai'i's cultural heritage and natural history, as well as its ancestral cultures throughout the Pacific.
www.bishopmuseum.org

CAPER - Center for Astronomy & Physics Education Research
With research centers in Laramie, Phoenix and Hilo, CAPER Team scholars develop innovative teaching strategies to improve learning of physics and astronomy.
http://CAPERTeam.com

Dietrich Varez
A beloved Hawai'i island artist who boldly traces the adventures and passions of a cast of mythical Hawaiian characters he has carefully researched in legend. He lovingly and faithfully depicts Hawaiians practicing the arts, skills, and values of Hawai'i.
www.dietrichvarez.com

Honolulu Museum of Art
Bringing together great art and people to create a more harmonious, adaptable, and enjoyable society in Hawai'i.
www.honolulumuseum.org

'Imiloa Astronomy Center of Hawai'i
"'Imiloa is a place of life-long learning where the power of Hawai'i's cultural traditions, its legacy of exploration and the wonders of astronomy come together to provide inspiration and hope for generations."
www.ImiloaHawaii.org

Journey Through the Universe: Program with the National Center for Earth and Space Science Education
A national science education initiative that engages entire communities -- students, teachers, families, and the public -- using education programs in the Earth and space sciences, and space exploration to inspire and educate.
www.journeythroughtheuniverse.org

Lyman Museum and Mission House
"To tell the story of Hawai'i, its islands, and its people."
www.lymanmuseum.org

PISCES - The Pacific International Space Center for Exploration Systems
A Hawaii State Government Aerospace Agency located in Hilo, Hawai'i that conducts environmentally safe tests on Hawai'i's volcanic terrain to validate advanced space technologies
www.pacificspacecenter.com

Polynesian Voyaging Society
Founded on a legacy of Pacific Ocean exploration, the Polynesian Voyaging Society seeks to perpetuate the art and science of traditional Polynesian voyaging the spirit of exploration. Through experiential educational programs they inspire students and their communities to respect and care for themselves, each other, and their natural and cultural environments.
www.pvs-hawaii.com

University of Hawaii at Hilo Department of Physics and Astronomy
UHH offers an excellent undergraduate program designed to prepare students for a wide range of careers in physics, astronomy and other sciences.
www.astro.uhh.hawaii.edu

University of Hawaii at Hilo Hawaiian Studies Program
Focused on a Hawaiian-based Cultural continuum that perpetuates the Hawaiian culture within a Hawaiian Language context.
hilo.hawaii.edu/catalog/hawaiian-studies.html
Ka Haka 'Ula O Ke'elikolani, College of Hawaiian Language
www.olelo.hawaii.edu

University of Hawaii Manoa - Hawai'inuiakea School of Hawaiian Knowledge (HSHK)
With respect and reverence for our ancestors, the mission of HSHK is to pursue, perpetuate, research, and revitalize all areas and forms of Hawaiian knowledge.
www.manoa.hawaii.edu/hshk

University of Hawaii Manoa - Institute for Astronomy
Explore the origin and nature of the Universe, develop new technologies for use in ground-based and space-based observatories and spread the understanding of astronomy.
www.ifa.hawaii.edu

Volcano Art Center
VAC is an educational organization founded in 1974 by a band of eclectic and energetic artists that continues to operate a successful fine arts gallery showcasing over 300 local artists in Hawai'i Volcanoes National Park.
www.volcanoartcenter.org

Books published by Petroglyph Press

A CONCISE HISTORY OF THE HAWAIIAN ISLANDS
by Phil K. Barnes, Ph.D.
HAWAIIAN LEGENDS OF VOLCANOES
by W.D. Westervelt, Illustrated by Dietrich Varez
HILO LEGENDS
by Frances Reed
HINA - THE GODDESS
by Dietrich Varez
I'A, SEALIFE COLORING BOOK
by Jan Moon
'IWA, THE HAWAIIAN LEGEND
by Dietrich Varez
JOYS OF HAWAIIAN COOKING
by Martin & Judy Beeman
KANEHUNAMOKU
by Dietrich Varez
KONA LEGENDS
by Eliza D. Maguire
LEGENDS OF MAUI
by W.D. Westervelt, Illustrated by Dietrich Varez
PELE AND HI'IAKA, A Tale of Two Sisters
by Dietrich Varez
PELE, VOLCANO GODDESS OF HAWAI'I
by Likeke R. McBride, Illustrated by Dietrich Varez
PETROGLYPHS OF HAWAI'I
by Likeke R. McBride
PLANTS OF HAWAII - HOW TO GROW THEM
by Fortunato Teho
PRACTICAL FOLK MEDICINE OF HAWAI'I
by Likeke R. McBride
STARS OVER HAWAI'I
by E. H. Bryan, Jr. and Richard Crowe, Ph.D.
THE STORY OF LAUHALA
by Edna W. Stall
TROPICAL ORGANIC GARDENING, HAWAIIAN STYLE
by Richard L. Stevens